AGRICULTURAL GEOGRAPHY : ISSUES AND APPLICATIONS

Agricultural Geography

Issues and Applications

Dr. R. S. DUBE

Gian Publishing House

29/6 SHAKTI NAGAR • DELHI-110007 • INDIA

GIAN PUBLISHING HOUSE
29/6, Shakti Nagar, Delhi-110007

ISBN 81—212—0083—0

PRINTED IN INDIA

Published by Mrs. Gayatri Garg for Gian Publishing House, Delhi and printed at S.C. Sharma Printers, B-III, H-146, Vishnu Garden, New Delhi-110018.

Dedicated to
the pious memory of
my reverend Guru,
Professor S.M. Ali,
the doyen of Indian geographers,
who made me what I am.

PREFACE

Modern Geography is in a state of rapid progress. Although the division between the 'old' and the 'new' geography is not very sharp, the modern geography has made so remarkable developments during the past few decades that it has now assumed the status of a real scientific discipline. It has, in fact, evidenced marked changes in its nature and scope with the arrival of a large body of contributions embracing a wide spectrum of ideas and opinions of geographers from different parts of the world.

It may be pointed out, however, that the attention of professional geographers has remained focused largely on the analysis of different facets of spatial dimensions of geographical phenomena. The traditional methods and tools of empirical observations and field studies have been relegated to a subordinate position owing to an excessive import of numerous quantitative techniques and many other ideas from sister disciplines of physical and social sciences. Quantification of spatial measurements and territorial relationships have become such a widespread geographical fashion today that the real nature of geography has considerably suffered on account of the consistent intrusion of methodological issues and ideas borrowed from these disciplines. Though the inherent strength of geography could resist such technical invasions and it has emerged now as a stronger discipline, the principal focus of geographical researches still remains centred on spatial dimensions and spatial problems concerning the welfare of mankind.

The geographers are yet to appreciate that their main professional object is the formulation of theories on the basis of empirical observations and analyses of variables occuring in different geographical situations. The identification of multifarious socio-economic and territorial dimensions of geographical facts has so far invited greater attention of geographers. Quite a substantial part of the academic energy of geographers has remained involved in converting

Geography into the so-called discipline of Applied Geography without realizing the fact that the basic task of geographical studies has always been the application of geographical methods for understanding of various regional and sectorial problems of different areas of the earth. The result of the academic chaos that has kept overcast the nature and scope of the science of geography for last several decades has resulted in geographers' inability to prove their worth either as regional scientists or as potential theoreticians of geographical events. This is evident from the fact that geographical studies have very rarely received significance in the decision-making process in the socio-economic planning of areas and that many theoretical pronouncements presently in vogue in geography are those which were enunciated by non-geographers. I am of the opinion that geographers must concentrate more on thinking how radical changes in geography may be brought about to validate greater relevance of geography as a regional science.

An effort has been made in this book to demonstrate how geographical researches may be utilised for framing theoretical postulates in order to widen the academic horizons of the subject. The present book is much different from the existing textbooks on Agricultural Geography. It includes thirteen essays covering a wide array of concerns in Agricultural Geography ranging from the problems related to the classification and mapping of data to those related to sufficiency and marketability of agricultural produces. The principal concern of this indeavour has been the identification of some of the outstanding issues in the field of Agricultural Geography and their application at different territorial scales. The pivot around which ideas and opinions revolve in this book is the theory on land-use cycle. In it the need for a greater appreciation of the significance of historical dimensions of the dynamism of agricultural phenomena has been stressed.

The Department of Geography at Dr. H.S. Gour Vishwavidyalaya, Sagar, India has served as an excellent base for carrying out my research activities for more than 25 years, the results of which have made possible the formulation of conceptual and methodological ideas presented in this book. I am immensely indebted to the principal architect of this Department, the late Professor S.M. Ali, whose high ideals and high standards of academic life have served as a constant source of inspiration to his numerous students and colleagues spread all over the country, of whom I have the honour of being one. I am grateful to Professor V.C. Misra, the present Head of the Department and Rector of this University, whose guidance and blessings have been responsible for the success of my research activities at Sagar. I also place on record my thanks to my colleagues and research students for lively discussions on several occasions. My thanks are also due to the following for permitting me to incorporate the copyright materials of my articles published by them : The Hong Kong Geographical Association; National Association of Geographers, India; Editor, *Geographical Outlook*. Department of Geography, Ranchi University, India; Institute of Indian Geographers, Poona, India; The Indian Geographical Society, Madras; and The Geographical Society of India, Calcutta.

Sagar, 1986

R.S. DUBE

CONTENTS

CHAPTER ONE

INTRODUCTION

Agriculture is the mainstay of human life as a principal source of foodstuffs and industrial raw materials. It is the most widespread and dominant economic activity of a greater segment of the world population, notwithstanding the fact that industrialisation and urbanisation have brought about substantial reduction in the relative significance of cultivation in the more developed parts of the world. In the less developed countries, the agricultural workers share about 60 percent of the total economically active population. The percentage of agricultural population is, in fact, much more higher in many individual African and Asian countries like Afghanistan (78 per cent), Bangladesh (83 per cent), Bhutan (93 per cent), India (63 per cent), Kampuchea (74 per cent), Lao (74 per cent), Nepal (93 per cent) and Vietnam (71 per cent), where agricultural workers constitute a very high segment of the working population.

The world population has evidenced rapid growth in the modern period. The staggering increase of population continues even today in many parts of the world. Demand for foodstuffs and industrial raw materials have-had a consistent uptrend during the whole course of modern period. Famines, starvation, malnutrition and undernourishment have been causing multifarious health problems in different parts of the world. It is not surprising, therefore, that geographers have focused their attention on the study of various aspects of agriculture the man's husbandry of land.

The principal objective of the study of the agricultural geography is to describe and explain the salient features of the agricultural landscape of different parts of the world. The agricultural landscape of a territory is the result of the existing infrastructural features of its agriculture, the agricultural processes that are in operation there, and of the stage at which its agricultural system has reached. The study of infrastructural aspects of agriculture and agricultural processes and their dependence on multifarious geographical factors have been the subject of geographical analysis and interpretation for quite a longtime. But features of agricultural systems have been very rarely examined in the context of their historical consequence in different regional studies of agriculture. It must be appreciated that salient features of the agriculture geography of a territory are intimately related with the stage at which its agricultural system has arrived at. The present monograph is an attempt at compiling some essays with a view to highlighting the significance of historical component of agricultural evolution over other determinants of agricultural system.

An observation of spatial variation of land use systems in different earth regions suggests a systematic succession of land-utilisation from the least developed to the most developed areas of the world. At the least developed extreme are the forest fallows associated with sparse population, shifting cultivation and primitive agriculture. At the most developed extreme are the multi-cropping systems associated with dense population, multiple cropping, improved modern agricultural techniques and high farm productivity. Between these extremes are found the bush fallow cultivation, short fallow cultivation and annual cropping systems arranged in a gradual succession towards the most developed parts of the world. Their dominant infrastructural characteristics and productive systems have shown close relation with the stage of the evolution of their agricultural systems. It must be appreciated, therefore, that spatial analysis of farm enterprises and farm operations would find explanation and meaning only when they are viewed in the historical context of the agricultural systems of which they are parts.

The following chapters in this monograph highlight some of the fundamental issues, the applications of which have been attempted at different regional scales in different sections of different chapters.

Geographers have concentrated their greater attention so far on the problems related with the present dimensions of agricultural land use and agricultural productivity. This may be easily corroborated from a large number of publications on various aspects of these areas of agricultural geography of different earth regions. The emphasis of geographical studies has been mainly on the analysis of present spatial and infrastructural configuration of agricultural activity with the aid of many advanced cartographic and statistical jugglery. These efforts have not gone waste. They have helped highlighting present problems with a view to identify future strategies and future priorities in agricultural practices. It is a pity that geographers have subordinated their role only as aides to the policy makers and planning agencies only. We cannot dispute the role of Geography as a welfare science and there is no harm if geographers work to stimulate the planning agencies which are directly responsible for the socio-economic upbringing of the human community. But geographers are not supposed to serve simply as facts compiling agencies and regional analysis of empirical observations. It must be appreciated that their principal task is the construction of theories regarding spatial and sectorial behaviour of geographical phenomena based on empirical observations for different points of time and for different spatial situations. Unfortunately, geographers have mostly failed to discharge their responsibility in this matter. This would be evident from the fact that most of the theories on agriculture have come from non-geographers. A casual look at the list of publications on different aspects of agricultural geography would also substantiate this observation. The issues raised and their applications attempted in the following sections of this book are humble efforts to theorise some of the outstanding issues of this branch of geography.

The failure of geographers to attend adequately some of the fundamental problems of handling of agricultural data has been highlighted in chapter two on classification and mapping of data. It would seem ridiculous that geographers have not been able to evolve consensus on a uniform scheme of classification and mapping system for agricultural data which could have accorded the regional studies greater relevance and coherence. This problem has, therefore, been highlighted and suggestions for a uniform adoption of data classification and mapping have been advanced in this chapter.

The land use studies have kept involved most of our geographical enterprise in this country for last several decades. But little theoretical results could be achieved except the magnification of problems related with agricultural efficiency and agricultural productivity. It has never been realised that agricultural efficiency and agricultural intensity in a territory are the outcome of the stage of agricultural evolution which that territory has arrived at. This issue has been highlighted in third chapter on land use cycle where effort has been made to present a theory of a Cyclic progression of land use systems. The issues raised in this chapter have been applied in the Madhya Pradesh in the chapters four to six. These chapters examine the validity of historical sequence of following, changes in farming technology and cropping intensity in Madhya Pradesh in order to testify the validity of the ideas of Boserup and Malthus. It would be stimulating to note that Boserupean and Malthusian conjectures do not find complete applications in the agrarian setting of this part of the world.

The subsequent sections of this volume are related with the issues of agricultural productivity, agricultural sufficiency, land utilization and over crowding, and with the problems of circulation of agricultural products from surplus to the deficit parts of a territory. It would be seen that these sections highlight the dominant issues which have been applied at different regional scales. The chapter seven provides the problem of estimating agricultural productivity and the subsequent chapter presents the application of productivity indices for the agriculture of Madhya Pradesh with a view to provide empirical laws regarding the spatial configurations of agricultural productivity.

The methods for evaluation of food sufficiency have been presented in chapter nine. It will be seen that an effort has been made in this chapter to present a modified approach to the idea of measuring food sufficiency in a territory. The ideas highlighted in this chapter have been used for estimating food sufficiency at various regional scales in chapters ten, eleven and twelve. The problem of distribution of land use system in an over-crowded rural environment is presented in chapter thirteen of this volume.

The problems of land use efficiency and food sufficiency eventually find expressions in the problems related with the circulation of agricultural products from surplus to deficit areas. This idea has provided basis for the content of chapter fourteen. The experience of the handling of the issues raised in this book have been concluded and attention of geographical community has been attracted to the realisation that academic chaos in agricultural geography must be allowed to settle down in the form of theoretical summarisation of past research enterprises.

Chapter Two

CLASSIFICATION AND MAPPING OF AGRICULTURAL DATA

One of the principal concerns of Agricultural Geography is the description and explanation of the spatial aspects of agriculture. The analysis and interpretation of spatial variations and spatial relationships of agricultural phenomena evolves classification of relevant agricultural data and their presentation into meaningful mapping devices. Though the geographical methodology has now evolved there seems little consensus on the problem of a rationally agreed procedure of classification of data for depicting them in distributional maps. The procedure adopted for grouping of data in different regional studies of agriculture is often varied and their classification is yet to conventionalise into some unanimously agreed basis. Personal discretion very often dominates the system of numerical taxonomy in agricultural studies and it seems desirable to examine the current classification procedures with a view to evolve some commonly agreed principles for data classification. In fact we should work to achieve a national consensus on the formulation of a scheme of classification of agricultural statistics in order to impart greater relevance to the study of spatial aspect of agricultural activities.

The essence of our enterprise in agricultural research is the formulation of hypotheses and empirical laws regarding the spatial distribution and spatial relationships of agricultural enterprises and farm operations. This may be accomplished from the observation of regularities in relationships which may be readily visualised in the inductive generalisations emerging on the maps showing spatial information of agricultural activities. Maps are, in fact, effective media for 'transmission' of areal data, which if classified on the basis of some sound grouping techniques can communicate the spatial behaviour of agriculture in a more efficient and meaningful form. In fact, the careful grouping of data is an inseparable cartographic exercise in agricultural studies, for a poor classification may produce misleading results. Currently a diverse set of classification procedures are in use for mapping agricultural data and a commonly agreed grouping procedure seems necessary.

We can appreciate the difficulty in laying down a precise system of classification of agricultural data, but their grouping on the basis of some carefully chosen common properties or relationships may easily fetch acceptance by the map makers. Since the groups are clusters of data, and the task of classification eventually involves agglomeration of individual values of a set of agricultural variable the grouping may be based on the following considerations :

(i) The number of classes to be chosen ; and

(ii) The determination of boundaries between the classes.

NUMBER OF CLASSES

The views of geographers and cartographers on the choice of number of classes for mapping are quite dividend. We can appreciate the difficulty in evolving unanimity on the adoption of a specific number of classes, for the choice would mainly depend upon the purpose of mapping. In certain situations even as small as two classes may be sufficient to reveal a desired pattern. For example, grouping of data into two classes may effectively bring out the surplus and deficit areas in terms of the sufficiency of farm output. The data in this case may be grouped into two types in the following manner : (i) those were food (per capita) production is more than the recommended requirement of energy for an individual, i.e. 2400 calories ; and (ii) those where food production is smaller than the recommended energy requirement of an individual. Adoption of a larger number of classes in this case may, though help in highlighting the level of food sufficiency and deficiency in different parts of the areas, not seem much desirable if the purpose of mapping is simply the identification of surplus-deficit pattern of food crop production.

In certain other situations, however, a greater number of classes may be desired to bring about the salient aspects of spatial configurations of different agricultural operations of an area. It is, therefore, difficult to prescribe a minimum or maximum number of classes, but the classes must usually not exceed five times the logarithm of the number of values to be covered (Davis, 1974). In selecting the classes, the primary consideration should be that the number of classes are neither too small to loose the salient details of spatial configuration, nor too large to allow several empty classes resulting in disintegration of sharpness in spatial patterns. The determination of class number should be based on the proper appreciation of the deficiency and sufficiency of small or large number of classes. Small number of classes, though loose details, facilitate ease in interpretation (Abler et al. 1972). The greater number of classes facilitate details at the expense of sharp spatial patterns. Ordinarily the classes must be not so many that the human eye finds it difficult to accommodate the patterns depicted on the map. The choice of class number, therefore, depends mainly upon the skill and discretion of the cartographer depending on the need and purpose for which the mapping is to be done. There may be situations, for example, where areas with extreme values may be desired to emerge on the map, as distinctively as any other type areas. This would necessitate adoption of a greater number of classes to allow separate classes for too high and too low values of the data set. In a certain other situation, however, the emergence of extreme areas may not be as relevant as identificaton of regularity and sharpness in the spatial patterns which would require selection of a smaller number of classes. Ordinarily, there may be as many classes as possible without empty classes and without impairing clarity and easy exhibition of patterns.

It may be concluded, therefore, that a universal consensus on the number of classes is neither feasible nor desirable. The issue will have to be left entirely at the discretion and skill of the map makes in accordance with the need and purpose of mapping.

The Class Limits

The choice of class interval is inseparable from the choice of class number. The class size depends mainly upon the number of classes used and the range of values to be covered. The class interval may be fixed or variable depending on the nature of data. Ordinarily, the interval may be fixed at equal spacing, where data on either side of the mean is equally distributed *i.e.* where data exhibit a normal distribution. But interval may vary at the geometric or any other spacing, where distribution of data is appreciable skewed.

Since basic purpose of classification is to classify the vital patterns of distribution of a data set, the class boundaries must be determined, in such a way that they can reveal the greatest amount of useful information about the events mapped. Groups are the clusters of values agglomerated on the basis of certain common characteristics. What must be the common parameters for permitting agglomeration of values into desirable number of classes ? Let us first examine the current grouping methods before attempting at the formulation of a procedure that may provide the yardstick for classification of data in all the regional studies.

Equal Classes

One of the approaches currently used for grouping of data is that where class boundaries are fixed at equal spacing by dividing the whole range of data into equal parts (Davis, 1975, p. 58) This is an easy and convenient procedure. Since the boundaries are arithmetically fixed in isolation of the nature of internal distribution of values this may evince misleading results. The method is, therefore, not much useful where scientific analyis and interpretation of patterns are to be attempted at.

Frequency Graphs

Another approach employed for the determination of classes is the use of simple scattergram or a cumulative frequency graph. The class boundaries in this method are located with the help of inherent breaks in the continuity of data. If the boundaries are determined entirely as exhibited by the breaks in the curve of the graph, the interval may be variable. But in practice the classes determined from the construction of the frequency graph are usually kept fixed at equal spacing. The method is considered quite useful, since it allows sufficient degree of flexibility in the determination of classes (Davis, 1975). But it fails to provide any standard yardstick for adoption of a uniform set of classification scheme for different regional studies. The classes determined by this method would be various for different regions and the method would, therefore, have little relevance for inter-regional comparisons.

Departure From a Typical Value. One of the best and scientific approach to the classification of data is the determination of degree of departure of individual values from a carefully chosen value (hereafter termed as typical value). The typical value is a point of reference in a data set from where the limits of different classes are measured and identified. Presently a number of methods are in use.

The *monothetic* classifications are based on the departure either from a central value—mean, median or mode—or from any other commonly accepted standard value. They may also be determined on the basis of some ranking techniques. The *polythetic* classification is based either on some such statistical procedures as average Z-scores (Dube, 1982) or on finding out some such composit scores for the variables covered as those suggested by Kendall (1939).

Departure from Centrality of Data : The classes may be worked out on the basis of departure of values from a central value—the mean, median or mode of the data to be classified. The departure may be worked out in the form of the per cent variation (Sharma, 1980), standard deviation (Ayyar and Dube, 1968), or Z-scores (Dube, 1982). The class limits determined by this method will be invariably spaced and arithmetically located. The method would, therefore, automatically remove subjectivity and flexibility in the determination of classes. Since this approach is based on the delimitation of class boundaries from the central value of a data, set, the classes will lack inter-regional comparability. The methods would, thus, have little relevance in regard to the formulation of a standard procedure that can impart inter-regional comparability in the spatial patterns.

Departure From A Standard Value : In some cases the Central Value does not serve much useful purpose in classification of data. For example, if regional growth rates of population are to be mapped, the central value will not have much relevance. A similar example that can be cited here pertains the mapping of deficit-surplus food producing areas where standard value of the food requirement of an average individual may be more usefully employed as a typical value for grouping of data. The classification made with reference to the regional mean would reveal the degree of variation from the regional average and would, therefore, impaire inter-regional comparability. But the classes worked out from a standard value would automatically standardise the grouping procedure to facilitate, comparability in the spatial patterns of different regions.

Polythetic Classification : The use of polythetic grouping becomes necessary where regional mapping is based on the composit patterns of a number of variables. A variety of methods are employed for the determination of such classes based on the desired number of variables. One of the old methods is the ranking based on the ranking of the sum of the ranks occupied by an area in respect of the variables covered (Kendall, 1939). Similar other techniques have been evolved to allocate composit scores to the areas which may be mapped in the ranked form. One of the useful methods is the determination of average Z-scores (Dube, 1982). The polythetic approach is an efficient way of presenting the multivariate picture of attributes. Since the allocation of ranks for the areas is based on intra-regional system of ranking, this method hardly finds much relevance for inter-regional comparability and thus lacks any standardised system of regional mapping.

Need to Standardise Classification Schemes

The study of agriculture of an earth region must be viewed as a part of the whole – the macro region, the state, the nation or the whole earth. The mapping of agricultural data at the micro scale must, therefore, be done in the manner that the maps of different regions may combine to form the map for the macro region similar to the practice adopted in surveying. Presently emphasis in the regional mapping is on highlighting intra-regional variability at the expense of ensuring inter-regional comparability. The problem of mapping and classification of data at the micro scale must, in fact, be inseparable from problem at the macro scale. The grouping of data at the micro scale must, therefore, be based on such a uniformly adopted classification procedure which may eventually enable the comparison and combination of different regional maps of the macro region. This can be conveniently accomplished if a common typical value may be identified for determination of classes. If a common typical value can be recognised, the classification scheme for all the areas of a macro region may be kept comparable to facilitate inter-regional comparability of patterns and the combination of different maps into one for the whole macro region. What can, then, be a common typical value ? In case of data for which some such standard values as needed for the mapping of food availability are available the grouping may be done on the basis of variation from that value. But what should be done in case of the classification of such data for which no such standard points are identifiable? The present author proposes the adoption of the national averages of the relevant variables as typical values for determination of classes. Many of the geographical variables mapped at the micro-scale derive meaning and significance when they are compared with national average values. For example, if mapping of area under rice in the districts of Madhya Pradesh is to be undertaken, the national average would provide the best point of reference from where the class boundaries of data in different regions of the country may be delimited. It is further proposed that the class limits may be fixed in the form of per cent variation from the typical value identified for a particular variable and the variation must be invariably measured in the multiple of five or any other unanimously accepted number. This would mean to say that the typical value will be treated as 100 and the boundaries of the groups on its both the sides will be determined in the form of the percentage variation in the range, for example, of 105. 110, 115 etc. on the upper side and 95, 90, 85 etc. on the lower side. The procedure may effectively contribute to the adoption of a uniform set of classification scheme, thereby standardising the whole mapping system in Agricultural Geography. It, therefore, follows that inter-regional comparability of patterns in the micro scale mapping and the ultimate combinations of regional maps can be safely ensured without prejudice to the quality of mapping system, if data of the relevant geographical variables are classified with reference to such typical values as a unanimously recognised standard value or the national average value of the relevant variable. For identification of such typical values is appended in Tables 1 and 2.

Exceptions

There may be, however, some such attributes where neither national average is identifiable nor any other standard value is in sight. For example, if a drainage frequency or relative relief map of a micro region is to be prepared, the determination of the typical value will be a difficult task. The present author proposes that in such cases the class boundaries should be determined neither in the form of per cent variation from the regional mean nor by dividing the range of values into equidistant parts if uniformity in the mapping system is to be ensured. The classes may be arbitrarily fixed in the multiples of five or any other unanimously recognised value to ensure uniformity in all the regional maps. There is nothing wrong in doing so. The tendency to base the classification essentially on a statistical method should not be over emphasised. The procedure of classification may vary depending on the nature of data and purpose of the study.

THE CHOICE OF MAPPING TECHNIQUE

The spatial distribution of agricultural data may be depicted with the help of a suitable mapping method. The choice of a mapping technique, however, depends primarily on the nature of data and on the purpose of mapping. If distribution of absolute numbers in isolation with their spatial relationships are to be depicted a simple pictorial or dot method may be found more useful. So may be the case for mapping of distribution of cows, irrigation pumps, wells, areas of rice cultivation and the like. Conversely, in case of mapping of qualitative data where distributional patterns are measured for some unit areas like tahsils in relation to some other geographical variables, the choropleth method may be found more useful for highlighting the salient aspects of the spatial configuration of the agricultural data to be mapped. If per hectare yield of different crops, for instance, are to be mapped, the choropleth system may prove an effective media of depicting their areal variations. If mapping is desired for the data available for different points in the space isopleth system would be more useful.

The use gravity models may be found useful in the analysis of spatial relationships of agricultural features of a territory. The gravity models have been more gainfully employed for the study of spatial mobility patterns. It may be emphasised that spatial variations of farm enterprises and farm operations are, to a great extent, result of the spatial transmission and spatial diffusion of agricultural knowledge. As such use of gravity model technique may be found useful in the study of agricultural geography of a territory.

BASE UNITS FOR MAPPING

The mapping in the regional studies is often done by selecting higher administrative units of the territory under study. The choice of base units seem to be variable depending upon the size of the study area. While selecting a base unit for mapping of agricultural elements, there are three fundamental considerations :

(i) the number of units should be large enough to reveal the spatial patterns effectively;

(ii) the periodic comparability of data for the base units should be possible; and

(iii) the data for these units for different points of time should be available.

The base units for mapping in agricultural studies should be selected in such a way that salient elements of the spatial configurations clearly emerge on the maps. There are three considerations in this regard :

(i) the areal size of these units should not be too small on the map to make the task of portrayal of patterns difficult. If these units are too small on the map, it would be difficult to lay different choropleths on them ;

(ii) their number should be large enough to impart effective exhibition of spatial patterns; and

(iii) these units have not undergone frequent boundary changes in the past to retard the temporal comparability of data.

It may be seen that the village affords an excellent unit for analysis and mapping of agricultural attributes for a number of reasons. Firstly, it is a fundamental unit of social organisation, material culture and agricultural holdings. Secondly, mapping of data at village level may exhibit the ultimate details of the space regarding the variable mapped. Thirdly, the village boundaries have stagnated for centuries to keep their data comparable for different points of time, while boundaries of higher administrative units have been very rarely stable during the past periods.

It may be remembered that mapping of data at the village level is feasible only for smaller earth regions. The village units of larger regions are not in the mappable size. It is suggested that village statistics may in that case be totalled to the higher administrative unit-patwari circle, revenue inspector circle, tahsil or district so as to keep the data at comparable level for different points of time.

It may be emphasised, however, that the number of base units should not be too large to yield excessive heterogenity of spatial patterns and make the task of spatial generalisation difficult.

NEED TO STANDARDISE CHOROPLETHING SYSTEM

Once we agree on the adoption of a uniform scheme of data classification for regional mapping we may work further to evolve a uniform set of choropleth system for preparation of regional maps. Presently the shading grades are employed to highlight the intra-regionly variations only and the areas with high or low values are usually allocated some such extreme choropleth shades as if still greater extremes in other parts of the macro region are not possible. The choropleth systems may also be used in an agreed form so that they become ultimately conventionalised in the manner the symbols on toposheet or weather maps have become.

The present author proposes that line shades may be allocated for the groups lying on the higher side of the typical value, while dot shades may be contained in the areas lying on the lower side of it. Different grades of line and dot shades may be allocated for the groups enclaved by specific per cent limits. An effort is made by the author of this paper to specify a set of shading patterns for different situations with the caution that these proposals are entirely provisional and further effort may be made to evolve a rationally agreed consensus on determination of a set of choropleth systems.

References

Abler, R. Adams, J.S. and Gound, P. (1972) : Spatial Organisation : The Geographer's View of the World. London : Prentice Hall.

Ayyar, N.P. and Dube, R.S. (1968) : Some Aspects of the Rural Population of Madhya Pradesh Geog. Out., Vol. 5.

Davis, P. (1975) : Science in Geography Pt. 3 : Data Description and Presentation, London : Oxford University Press.

Dube, R.S. (1982) : Social Well-Being in Madhya Pradesh. Trans. Ind. Geog. Vol. 4 (2).

Kendall, M.G. (1939) : The Geographic Distribution of Crop Productivity Jour. Royal Stat. Soc. Vol. 162.

Dube, R.S. (1979) : Population of Rewa Plateau : Kanpur : Sahitya Ratnalaya,

Chapter Three

THE THEORY OF LAND USE CYCLE

All terrestrial phenomena seem to change in a cyclic sequence. A careful observation of the patterns of changes evinced by different geopraphical elements would suggest that they follow cyclic rythms in the course of their evolution and development. Some cycles are so well documented and substantiated (e.g. normal cycle of erosion, climatic demographic cycle) that their cyclic progression is now well established in geographical literature. Other geographical elements also seem changing in a cyclic fashion during the course of their periodic change and growth. It seems that nature has prescribed a precise course of growth to all the terrestrial elements. The growth of land use system of a region does not seem to be an exception to this corollary of change occurring in the earth space. But it has attracted little attention of the agricultural scientists, economists and geographers and its theoretic frame is yet to find space in geographical research.

The evidences on the past land use systems are quite fragmentary. Available information, however, suggest that there was a gradual shortening of fallow in western Europe during and after the Middle Ages ending in the change to annual cropping in the second half of the 18th century followed by the present phase of multi-cropping system (Boserup, 1965, pp. 16-17). European evidences inform the process of fallow shortening ever since the neolithic period (Steward, 1956) and archaeological and historical evidences combine to inform that Europe had evidenced successive change from neolithic forest fallow to systems of shifting cultivation on the bush and grass land followed first by short fallow system and then by annual and multi-cropping systems in the recent times.

If we could accept the view that present is key to the past, a definite sequence of the past land use systems of the world may be identified from the observation of spatial sequence of change from least developed part to most developed part of the earth. It would be seen that at the one extreme of the present land use systems of the world are those parts of the earth where remnants of the earliest form of land use systems are practised. Here one or two years period of cultivation is followed by a fallow period of at least twenty years. These systems of cultivation are associated with sparse population, primitive agricultural techniques and least intensity of cultivation. At the other extreme are those parts of the world where most of the land is put under cultivation and where multi-cropping system is practised. They are associated with dense population, modern farm technology and high productive intensity of land. Between these extremes of the spatial configuration of the present world land-use are arranged the land use systems of intermediate intensities into well defined sequence of the bush-fallow cultivation, short-fallow cultivation and annual cropping systems demonstrating the order and direction of change that must have occurred in the growth of past land use systems in the world.

If spatial configurations of the present land use systems may be accepted as any yardstick of the sequence of change in the past land use systems, the pattern of their growth may be reconstructed from the observation of the distributive behaviour of different land use classes in any part of the earth. This will be a helpful exercise, since recorded data of the past land use patterns are indeed fragmentary and speculative. It would be better to depend on the present facts, rather then wandering into the darkness of the past speculative data.

The basic consideration in the identification of the sequence of growth of land use systems in a region is the determination of the degree of intensity of production. This is often done with the help of the trend of change in the proportion of fallows and culturable wastes which usually assume a declining trend with the forward progression of changes in the land use systems and with that of the evaluation of degree of agricultural intensity in a region Boserup (1965, p. 16) prefers to adopt the following sequence in the order of increasing intensity :

(i) Forest fallow cultivation;

(ii) Bush-fallow cultivation;

(iii) Short-fallow cultivation;

(iv) Annual cropping; and

(v) Multi-cropping. Evidently, the sequence of change in a region will be from the forest-fallow to multi-cropping systems. It may be, however, pointed out here that the present theories on the evolution of land use systems in a region are based on the idea of a forward progression of agricultural intensity and a retrograde progression has never been envisaged. Certain facts of the present land use systems found in different parts of Madhya Pradesh suggest that a retrograde progression in the growth of the land use systems is possible to accord cyclic course of progression in the growth of land use in a region.

EXPERIENCES FROM MADHYA PRADESH

The agricultural growth in Madhya Pradesh—the central Indian State, is presently passing through a transitional phase where a mixed pattern of short-fallow cultivation, annual cropping and multi-cropping systems in association with diminishing culturable wastes is practised. A little less than one-tenth of the total geographical area of the state is classed and under the short-fallows and the culturable wastes. The long-fallow cultivation is practised as remnant of the primitive phase of land use systems in the tribal forest enclaves of the State. They have historical relevance only, since they do not form extensive stretches in the State. The current and old fallows occupy about 4.56 per cent and the wastelands about 4.31 per cent of the total area of the State. Though the incidence of the fallows and the wastes is substantially lower than the national average, the trend of spatial and temporal changes exhibits such a peculiar pattern that it deserves attention of the geographers.

GROWTH OF FALLOWING

Temporal Trends

The pattern of temporal changes in the proportion of the fallows and the culturable wastes in Madhya Pradesh has been examined in more details in chapter four of this monograph. It would be seen that share of fallows has recently increased after 1972 after a long period of their reduction. Fallowing activity in Madhya Pradesh presents a perplexing situation, since such a progression in the land use systems developing under the influence of sustained population growth is difficult to occur and has so far remained unknown in the history of the growth of land use system in the world.

The population of the State has been increasing rapidly ever since the beginning of the period under review, viz. 1956-57. The population has, in fact, doubled during the period 1951-81, notwithstanding the fact that the cropped area has increased barely by 120 per cent. If we regard the process of agricultural change in the primitive communities, like those living in Madhya Pradesh, as an adaptation to gradually increasing population densities (Boserup, 1965, p. 18), the proportion of the wastes and the fallows must attain a declining trend, since growing population pressure on the crop land would necessitate expansion of cultivated area as part of farmer's initial enterprise to inhance agricultural production. The declining trend in the proportion of the Wastes is an adequate manifestation of the dynamic approach to the change in land use systems, but retrograde trend in the change of fallow cultivation in Madhya Pradesh invites adequate explanation.

The inverted growth of the fallow cultivation in Madhya Pradesh may find a plausible explanation in the decrease of soil fertility caused by shifting fallow cultivation to the annual and multicropping systems under the influence of sustained demographic growth as pronounced by Boserup (1965). It has been emphasised (Boserup, 1965) that in the initial phases of agricultural growth when population is sparse, the soil fertility is preserved by rotation of a system of fallow cultivation. But as densities and the resultant population pressures increase, the rotational system of fallow cultivation becomes difficult to continue for want of sufficient agricultural land and excessive use of the cropland leads to the exhaustion of soil fertility unless a quick introduction of agricultural intensification is allowed to come about. The demographic influences did operate to bring about gradual dimunition of the wastes and the fallows in Madhya Pradesh as both of them evidenced diminishing proportion, but the process of agricultural intensification does not seem to have come about as quickly as decrease of the wastes and fallows and the increase in population densities have occurred. Such a trend is bound to cause rapid exhaustion of soil fertility that would eventually lead to the reversal in the fallow cultivation as measure for stimulation of soil fertility. This is substantiated from the study of the temporal changes in cropping intensity and yields of the crops envolved in fallow cultivation in the State.

The Trend of Cropping Intensity

The temporal changes in the cropping intensity in Madhya Pradesh are discussed in details in chapter six. Though, the proportion of multi-cropped area has increased from 11.8 per cent of the total net sown area in 1957-58 to 14.3 per cent in 1979-80, registering a change of barely 121 per cent during the 25 years period the operation of crop intensification has not been sufficient to ensure adequate farm productivity for the rapidly growing population. The decrease in farm productivity is likely to compel the farmers to revive the previous system of fallow cultivation on their poorer farms, that would eventually bring about a retrograde trend in the temporal trend of the proportion fallows in the present system.

THE TREND OF CROP YIELDS

The temporal changes in the yield of crops envolved in fallow cultivation in Madhya Pradesh support the conjecture of gradual loss of soil fertility as a determinant of the inverted change in fallow cultivation. The principal crops envolved in fallow cultivation in the State include jowar, bajra, tur, kodon, kutki linseed, cotton, urd, moong, masoor and gram. The principal crops brought under the intensification programme include only rice, wheat and sugarcane. The trend of change in the yield of the fallow crops suggests that there has been little change in their yield rates. They have had fluctuating trends owing to the vagaries of weather conditions and the overall trends have been either stagnant or declining as detailed in chapter six. This would manifest the proposition that shift of land use from fallow cultivation to the annual or multi-cropping systems has made the farmers difficult to ensure stimulation of productivity of the poorer section of their farms. The development of such a situation would logically lead to the retrogression of the previous trend of declining fallow cultivation.

FOOD SUFFICIENCY TREND

Madhya Pradesh produced about 4.27 million tonnes of foodstuffs in 1951 giving a per capita daily availability of about 450 grams or 1585 calories of food energy. It was evidently at an acute deficient position, since its food supply was deficient by about 34 per cent. The subsequent agricultural growth was, however, able to register substantial increase in the farm productivity and the State became self sufficient in respect of its food production in 1961, when it produced about 887 grams or 2449 calories or per capita daily availability of food energy. The food sufficiency started deteriorating thereafter and in 1971 the per capita daily availability of food was 808 grams or 226 calories followed by 690 grams or 1665 calories in 1981. This would suggest the failure of agrarian growth to increase hand in hand with population growth.

It may, thus, be noted that land use system was able to stimulate the Boserupean system during 1951-61 in Madhya Pradesh and could conveniently supercede the pre-1951 Malthusian scarcity condition. The Boserupean system, however, again started debilitating in the post-1961 years due to the faster rate of population increase.

The Land Use Cycle

The preceding account of the dimensions of land use systems in Madhya Pradesh suggest a definite sequence of changes in the agricultural landscape of the State. We may conveniently envisage existence of hunting and gathering communities living in the paleolithic periods followed by a long period of forest fallow cultivation systems in the chalcolithic and neolithic periods when population was too small to stimulate any significant growth in land use (Dhavalikar and Possehl, 1974). This is evidenced from the archaeological evidences on domestication of animals and cultivation of plants gathered from the Mesolithic and neolithic sites in the neighbouring areas of the State (Sharma, 1980, p. 110.) The bush fallow and short-fallow cultivations must have followed subsequently, but we have no written records to substantiate our conjectural conclusions. The short-fallow cultivation system seems to have continued into the earlier parts of the 20th century. Some written accounts, however, inform the periods of sudden depletion of population due to famines, droughts, epidemic disasters and wars causing heavy destruction of labour force resulting into a complete reversal of land use systems. Many puranic and historical accounts may be quoted to substantiate incidence of such events. One of such disasters is stated to have occurred in the Malwa where Marathas invaded the territories with such a force that the population and habitations perished and the rich short-fallow cultivation system was completely destroyed to assure a retrograde progression leading to the forest fallow cultivation system.

With the establishment of an overall peaceful conditions in the British period and a gradual growth of population, a greater part of Madhya Pradesh seems to have passed into the short-fallow cultivation system until the advent of a period of rapid population growth after Independence of the country speeding up a quick dimunition of the fallows and the wastes and shift of the fallow cultivation to the annual and multi-cropping systems. In the areas where intansification process helped preservation of soil fertility, the annual or multi-cropping systems continued to be practised. So is the case in Bastar and Lashkar regions. But in the areas where intensification process lagged behind the excessive use of the crop-land under an annual or multi-cropping system, a retrograde progression of the land use bacame inevitable. So has happened in the Chhattisgarh Type land use system.

The occurrence of a retrograde progression in the land use systems and intermittent operation of Malthusian and Boserupean forces in the recent past in Madhya Pradesh would suggest that a forward progression of land use changes may be inverted if intensification processes fail to advance hand in hand with the land use dynamics brought about by sustained population growth. These may be termed as interruptions in forward progressions of land use changes since processes of demographic and land use dynamics would eventually advance further and overcome the situation and advance the changes towards the systems of higher intensities.

We may now advance our thinking to the phase that may eventually follow the phase of multi-cropping systems. It is likely that intensification of agriculture fails to sustain productivity and soil fertility deteriorates owing to variety of causes, e.g. environmental changes, depletion of labour force by wars, epidemic, emigrations and the like. If any such situation occurs the land use systems may adopt a retrograde progression and move backward to the initial phase of the growth of land use in a region. This would accord the course of land use changes a cyclic form to justify the relevance of the theory of land use cycle as advanced in this essay.

It may be emphasised here that the retrograde change in the land use system may be brought about either from the simple operation of Boserupean dynamism in land use systems, viz, the exhaustion of soil fertility or from the operation of Malthusianism, viz. excess of population resulting into the loss of labour from increasing death rate caused by famines, vices etc. It must be, however, remembered that both of these processes would be complementary to each other. The Boserupean situation would vigour at the process for Malthusianism to occur quicker and the Malthusianism would contribute to the retrogation of Boserupean dynamism to operate.

Conclusions

The growth of land use systems in Madhya Pradesh seems to have adopted retrograde progression in the past. This is inspormed from general conjectural estimates, historical records and present data on land use. The inverted sequence of growth in the land use systems of a region may suggest an eventual reversal of the land use systems after a long operation of multi-cropping system. The inverted growth of the land use may be brought about by exhaustion of soil fertility, environmental deterioration and loss of population force from emigration, famines, epidemics, wars etc. This may lead to the intermittent operation of Malthusian and Boserupean systems of land-people relationships.

References

Boserup, E. (1965) : The Conditions of Agricultural Growth. London Allen and Unwin.

Dhavalikar and Gregory, L. Possehl (1970) : "Subsistence Pattern of an Early Farming Community, Western India", Puratattva, No. 7, pp. 39-46.

Dube, R.S. (1986) : Population Pressure and Agrarian Change in Madhya Pradesh. New Delhi : Concept Publishing Co.

Dube, R.S. (1984) : "Agricultural Productivity in Madhya Pradesh", Annals Nat. Asso. Geogr. India, Vol. IV (2), pp. 13-24.

Dube, R.S. (1986): "The Anomalous Growth of Fallowing in Madhya Pradesh", Geog. Out look (Forthcoming).

Dube, R.S., : "Innovations in Farm Technology" Marketing Geographer, (Forthcoming).

Dube, R.S. et al. : "Intensity of Cropping in a Boserupean Setting" (Unpublished paper).

Misra, R.P. (1982) : Population and Food Supply in Madhya Pradesh. Unpublished Ph.D. Thesis, University of Saugar.

Sharma G.R. (1980) : History to Pre-history : Archaeology of the Ganga Valley and the Vindhyas, Kausambi Museum, Allahabad.

Stewart, O.E. (1956) : Fire As The First Great Force Employed by Mans in W.L. Thomas, Jr. (Ed.) Man's Role in Changing the Face of the Earth, Chicago, pp. 156-59.

Chapter Four

THE ANOMALOUS GROWTH OF FALLOWING IN MADHYA PRADESH

The duration and extent of the fallow land afford a reliable indication of the stage of agricultural growth in a territory. Boserup, the leading exponent of the dynamic theory of land use, while visualising the changes in land use systems as adaptations to the demographic dynamism, has shown a gradual shortening of the fallows with the progression of land use systems to the higher intensities.[1] It is emphasised that in the initial phases, when population is small, a system of long fallowing is practised with little input of capital and labour per unit area, since the long period helps to preserve soil fertility. But as the density of population increases the fallows are shortened and the soil fertility is difficult to preserve through the practice of long fallowing and it becomes necessary to introduce improved methods of farming. The fallowing phase consequently passes over to the annual cropping and multicropping systems. An examination of the fallowing activity in Madhya Pradesh follows in this Chapter in order to test the validity of Boserup vis-a-vis the growth of fallowing in a territory.

The Present Situation

The present land use in Madhya Pradesh is based predominantly on the annual cropping system. Though marginal remnants of the previous short-fallow system and the signs of marginal succession of annual cropping to the multi-cropping system are noticeable in the agricultural landscape of the state, the dominance of annual cropping is evidenced in the conspicuously low share of fallowing (envolving 7.9 per cent of the cultivable area) and multi-cropping (including 10.5 percent of the cultivable or 14.4 per cent of the net sown area) practice in different parts of the State.

The population of the State has been registering rapid growth for several decades. The population has, in fact, doubled within a fairly, short period of thirty years from 1951 to 1981. Such a staggering increment in population number is bound to bring about significant changes in the agricultural geography of the State.

The Anomalous Growth of Fallowing

The pattern of temporal changes in the incidence of fallows during the period 1956-1981 (for which reliable statistical information are available) is summarised in Table 4.1. It would be evident from the Table that share of fallows in the total geographical area of the state has gradually declined from 5.34 per cent in 1956-57 to 3.50 per cent in 1971-72 in accordance with the trend of change visualised by Boserup. The fallowing has, however, taken a retrograde trend thereafter as the incidence of fallows has evidenced gradual increase and they have expanded to approach in 1980-81 the proportion they shared in 1956-57. Though the length of period under review is too short for making any theoretical conclusion, the recent resurgence of fallowing activity in Madhya Pradesh makes the validity of the proposition advanced by the propogators of the dynamic theory dubious.

TABLE 4.1

Madhya Pradesh : Growth of Fallowing (1956-57 to 1980-81)

Years	*Fallows (Million hect.)*	*Years*	*Fallows (Million hectares)*
1956-57	2.36	1969-70	1.53
1957-58	2.71	1970-71	1.57
1958-59	2.40	1971-72	1.55
1959-60	2.42	1972-73	1.64
1960-61	2.20	1973-74	1.65
1961-62	2.04	1974-75	1.72
1962-63	2.02	1975-76	1.63
1963-64	1.96	1976-77	1.87
1964-65	1.78	1977-78	1.76
1965-66	2.01	1978-79	1.75
1966-67	2.08	1979-80	2.23
1967-68	1.77	1980-81	2.01
1968-69	1.63		

Source : Director of Agriculture, Madhya Pradesh, Agricultural Statistics, 1981.

The inversion of fallowing in Madhya Pradesh seems to be attributable to the rapid exhaustion of soil fertility resulting from the over use of the farms for cultivation unaided by sufficient technological innovations. A number of indicators would support this view.

The Diminishing Returns

The yields of the crops which are usually envolved in fallowing in this state (jowar, bajra, tur, kodon-kutki, linseed, cotton, urd, moong, masoor and gram) have registered little change since 1956-57. Though their yield have had fluctuating trends due to the vagaries of weather conditions, the overall trend of change has been either stagnant or declining (Table 4.2) suggesting that agricultural efficiency has not kept pace with the growing population and in some cases had even marked dimunition. If technological innovations fail to occur quickly, the preservation of soil fertility and farm productivity from the practice of fallowing is bound to revive, since areas of marginal productivity will be left out as fallow, in order to allow them stimulation of soil nutrients.

Insufficiency of Technological Innovations

The insufficiency of technological innovations on the farm may result in the rapid exhaustion of soil fertility for regular annual cropping operations on the farm unaided by improved farm technology would bring about dimunition of plant nutrients in the farm soils. In Madhya Pradesh the efforts for enhancing farm output has centred mainly on the horizontal expansion of the cultivated area, which is evident from the gradual decline in the area of fallows and culturable wastes. The degree of technological innovation are yet very low.[2] This is evident from the extremely low coverage of irrigational (Table 4.3), biological (Table 4.4) and mechanical (Table 4.5) inputs of agricultural innovations. As would be evident from the data appended in Tables 4.3 to 4.5 the farm technology is yet predominantly conservative and traditional. About 88.5 per cent of the total cropped area is yet unirrigated, and a greater part of the farming is yet handled by the traditional soil scratchers and other old farm implements. The application of high yielding seeds, fertilisers, and plant protection techniques envolved a very small proportion of the total crop land (Table 4.4). Such an insufficiency of the application of agricultural innovations is bound to cause rapid deterioration in

TABLE 4.2

Madhya Pradesh : Crop Yields (Kilogram/Hectare) 1957-1980 (3 years moving averages)

Years	*Jowar*	*Bajra*	*Kodonkutki*	*Gram*	*Tur*	*Masoor*	*Cotton*	*Linseed*
1957-58	677	546	210	557	712	359	309	182
1958-59	664	557	208	543	701	389	233	200
1959-60	661	561	242	798	801	429	237	224
1960-61	594	534	220	586	720	422	199	223
1961-62	627	582	199	582	688	373	219	218
1962-63	612	612	205	558	505	345	212	210
1963-64	744	722	208	560	539	343	268	203
1964-65	744	772	199	536	517	372	265	175
1965-66	727	786	154	494	499	336	232	179
1966-67	722	718	157	486	478	312	209	192
1967-68	698	592	183	471	480	313	221	208
1968-69	698	535	221	529	611	359	244	221
1969-70	627	541	220	522	668	383	226	214
1970-71	609	582	222	588	778	390	236	231
1971-72	670	598	217	631	860	391	240	234
1972-73	658	742	219	645	775	374	236	243
1973-74	754	572	210	625	763	387	260	255

1974-75	719	691	221	611	741	407	246	267
1975-76	686	579	194	594	736	418	274	238
1976-77	706	601	216	557	705	383	228	226
1977-78	708	623	197	541	650	360	229	208
1978-79	651	477	180	510	515	319	220	193
1979-80	658	495	162	535	568	342	221	187

Source : Directorate of Agriculture, Madhya Pradesh, Agricultural Statistics, 1981.

soil fertility that would necessitate, in its turn, the resurgence of traditional fallowing techniques for preservation of soil nutrients.

TABLE 4.3

Madhya Pradesh : Growth of Irrigated Area

Years	*Percentage of the cropped area*	*Years*	*Percentage of the cropped area*
1956-57	4.8	1969-70	7.2
1957-58	6.0	1970-71	7.4
1958-59	4.8	1971-72	8.2
1959-60	5.2	1972-73	8.5
1960-61	5.2	1973-74	8.2
1961-62	5.2	1974-75	8.3
1962-63	5.6	1975-76	8.9
1963-64	5.6	1976-77	9.8
1964-65	5.8	1977-78	10.4
1965-66	5.6	1978-79	11.1
1966-67	6.1	1979-80	10.7
1967-68	5.9	1980-81	11.5
1968-69	6.8		

Source : Directorate of Agriculture, *Ibid.*

TABLE 4.4

Madhya Pradesh : Coverage of Biological Innovation, 1981

Items	*% of the cropped area covered*
High Yielding Seeds	14.58
(*i*) Paddy	30.9
(*ii*) Wheat	29.7
(*iii*) Jowar	18.9
(*iv*) Maize	16.1
(*v*) Bajra	24.6
Plant Protection	
(*i*) Seed Treatment	5.68
(*ii*) Crop treatment	8.54
(*iii*) Rat control	2.27
(*iv*) Safe storage	11.05

Fertilizers	9.25 per hectare kg.
(i) Nitrogen	5.72
(ii) Phosphate	2.79
(iii) Potash	0.74

Source : Directorate of Agriculture, *Ibid.*

TABLE 4.5

Madhya Pradesh : Density of Farm Implements, 1981

Implements	*Density (No. per 1000 hectare of the cropped area)*
Wooden ploughs	232
Iron ploughs	8.6
Tractors	1.07
Bullock carts	1.11
Irrigation pumps	16.91
Expellers	0.55
Plant protection machinery	1.07
Persian wheels	3.96

Source : Directorate of Agriculture, Madhya Pradesh : Agricultural Statistics, 1981.

Sluggish Growth of Croping Intensity

The evidence of diminishing returns of labour productivity may be evidenced from the sluggish growth of multi-cropping practices. The share of multi-cropped area in the net sown area has registered only marginal increase from 11.8 per cent in 1956-57 to 14.3 per cent in 1980-81 (Table 4.6) giving an increase of barely 121 per cent during this period.[3] This would suggest that operation of agricultural intensification has not been sufficient to ensure adequate farm productivity. In the absence of the quick operation of intensification processes the farm soils will be rapidly depleted and the previous system of fallowing would become inevitable for stimulation of soil fertility.

TABLE 4.6

Madhya Pradesh : Cropping Intensity 1956-1972
(3 years moving averages)

Years	*Multicropped area (% of NSA)*	*Years*	*Multicropped area (% of NSA)*
1957-58	11.8	1969-70	11.1
1958-59	11.8	1970-71	12.1
1959-60	13.2	1971-72	12.4
1960-61	13.5	1972-73	13.2
1961-62	13.1	1973-74	12.4
1962-63	13.3	1974-75	13.1
1963-64	13.1	1975-76	12.5
1964-65	11.4	1976-77	13.7
1965-66	9.2	1977-78	14.1
1966-67	8.2	1978-79	14.3
1967-68	9.1	1979-80	14.3
1968-69	10.5		

The Spatial Ramifications

The share of fallows has registered marked spatial variations in Madhya Pradesh (Fig. 4.1). The percentage of fallows in the total geographical area of the district varies from 0.7 per cent in Sehore district to 11.7 per cent in Mandla district. The areas where incidence of fallows is greater than the national average of 6.8 per cent include about one-fifth of the total districts which form a compact stretch across the Satpura-Maikal region and in some areas of the Bundelkhand-Rewa uplands. The proportion of fallows gradually diminish on either sides of the Satpura-Maikal belt (Fig. 4.1).

The spatial dimensions of fallowing seem to give an intimate association with the nature of terrain, for hilly and dissected areas of the State support greater extent of the fallow. Though the application of dynamic theory of the growth of fallowing may have some serious

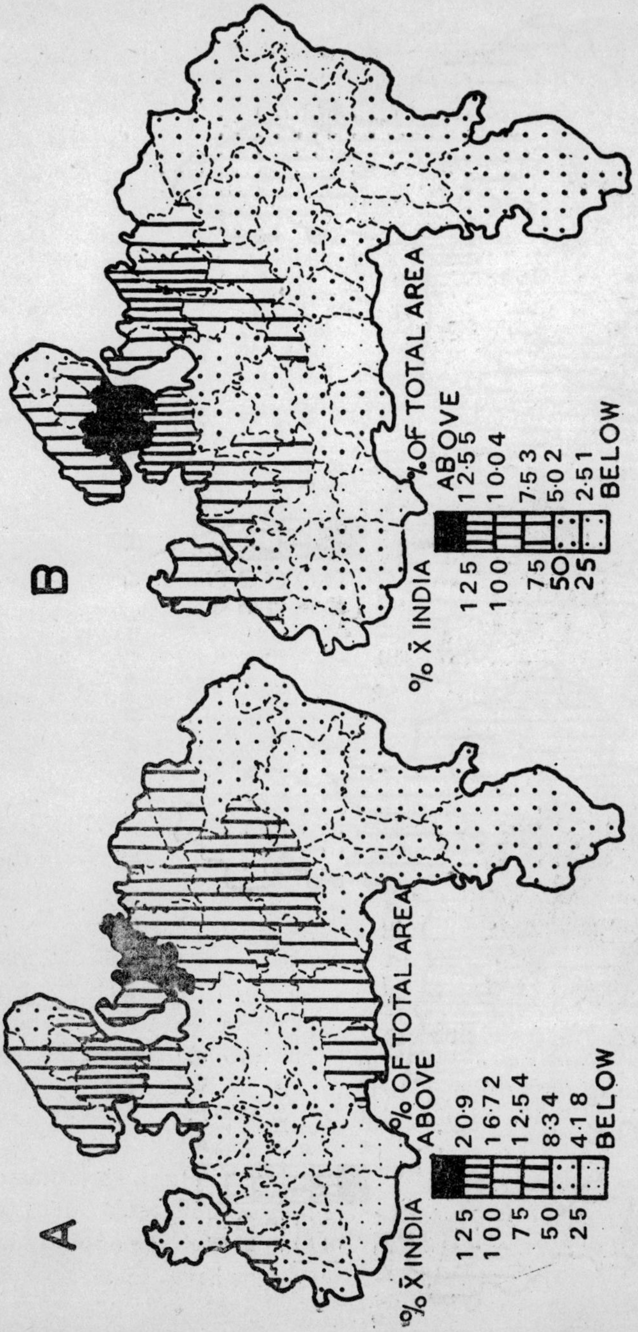

Map. 4.1: Madhya Pradesh A. Fallow and wastelands, 1981 B. Wastelands, 1981.

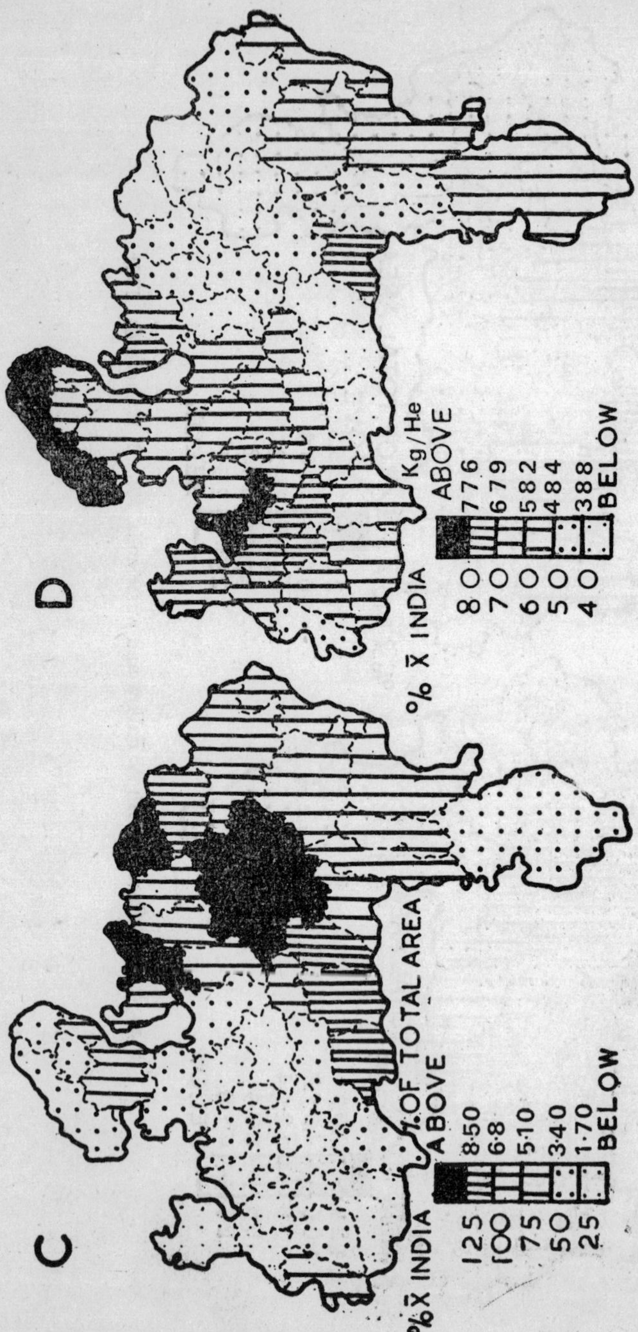

Map 4.1: Madhya Pradesh C. Fallow lands, 1981 D. Foodcrops : yield rate, 1981

difficulties since some more densely peopled parts of the State,[4] *e.g.* Jabalpur and Rewa, registered high percentage of the fallows, there is an evident sequental decline in the incidence of fallowing from the less peopled parts to the more peopled areas of the state. This would suggest that growth of fallowing activity had a sequential progression ever since the remote past and their remnants have been left out in the present spatial configuration of fallowing in the state.

The Modified Dynamic Theory of Growth of Fallowing

We do not have sufficient reliable data to identify the course of changes that may have occurred during the whole extent of the archaeological and historical past. The fragmentary evidences[5] would suggest that cultivation of soils and domestication of animals must have commenced in the neolithic period. The initial form of the cultivation must be the long forest-fallow cultivation succeeding subsequently to the bush fallow cultivation system. It is difficult to reconstruct a precise course of changes in the fallowing activity in the pre-historic and historic period, with the help of available information, but it may be easily surmised that the overall sequence of changes until the recent past must be similar to those propounded by Boserup.[6] This is substantiated from the spatial configuration of the land use systems arranged between the forest fallowing at one extreme and the multi-cropping at the other (Fig.4.1). It is also conceivable that population growth would stimulate these changes. It may, however, be pointed out that spatial associations of the land use systems and population concentration are difficult to surmised, since present land use system may determine the spatial variability of population distribution and vice versa. Further, investigations may be necessary to verify the demographic or physical adaptation of land use systems in the state.

We have documentary evidences to conclude about the sequence of changes that may have followed the short-fallow cultivation system in this state. It may be conveniently surmised that the increase in population density would contribute to the reduction of fallowing and introduction of annual cropping system, since demographic growth would necessitate extension of cultivation to the other cultivable areas. The sequence of events which would follow the annual cropping system would depend upon the efficiency of the land system to adopt technological innovations. If technological changes are swift to cope up with population growth, the annual cropping would give quick passage to the multi-cropping system. If technological innovation fail to proceed hand in hand with the increase in population density the annual cropping system would revert back to the previous fallowing systems, since soil fertility would need stimulation from the traditional system of rotation of fallows. It may be pointed out that this would lead to the failure of the Boserupean system of population-resource relations and Malthusianism would get upper land in producing conspicuous resource imbalances. This is in fact occurring presently in Madhya Pradesh.

The evidences of a retrograde growth of fallowing in Madhya Pradesh have serious implications vis-a-vis the theoretical pronouncements advanced by Boserup.[7] The Boserupean theory does not envisage the backward growth of land use system. The evidences in Madhya Pradesh, suggest that a retrograde progression in land use may result from a variety of factors e.g. exhaustion of soil fertility,[8] depletion of farm labour force,[9] destruction of population by epidemic, economic or political disorders.[10] It is possible that after a brief period of retrogration the land use systems may again adopt the forward trend of their progression. But what will heppen after the land use system has reached at its peak in the multi-cropping phase? Even if the population regulation succeeds to bring about zero growth in population change, the farm technology may fail to preserve for an indefinite period of time. The technology and the land use systems may start taking the inverted backgrowth. Malthusianism may have its way to the destruction of population and doomsday may bring a complete reversal in the land use systems. The cycle of land use may become complete.

Conclusions

The fallows have registered an inverted growth in Madhya Pradesh for some time. The resurgence of short-fallowing practices in a sustained demographic growth contributes to weaken the validity of the theory propounded by Boserup. The progression of land use in a territory may be interrupted by brief period of retrograde growth and the land use may ultimately adopt a backgrowth due to the failure of farm technology to preserve soil fertility or due to depletion of population from a complex set of socio-economic reasons. This would mean therefore that the land use growth may ultimately return to its initial phase to complete its cyclic progression. This would also mean that Malthusianism may eventually overtake the Boserupeanism.

Notes and References

1. Boserup, E. : Conditions of Agricultural Growth. London : Allen & Unwin, 1965.
2. Dube, R.S. : "Innovations in Farm Technology in Madhya Pradesh" Communicated to the Uttar Bharat Bhugol Patrika, Gorakhpur, 1984.
3. Dube, R.S. : "Intensity of Cropping in Madhya Pradesh" Communicated to The National Geographer, Allahabad.
4. Ayar, N.P. and Dube, R.S. : "Some Aspects of the Rural Population of Madhya Pradesh" Geographical Outlook.
5. Dhavalikar, M.K. and Gregory, L.P. : "Subsistence pattern of an early farming community of Westetn India", Puratattva, No. 7, 1974; Sharma, G.R. History to Pre-history, Kausambi Museum, Allahabad, 1980.
6. Boserup, E. : *op. cit.*
7. *Ibid.*
8. The existence of decreasing soil fertility has been elaborated in the present paper in its preceding sections.
9. There are evidences that farm labours are moving out from Chhattisgarh with the result that agricultural lands of marginal productivity are being left out as fallows and wastes.
10. Destruction of population may be caused by epidemic or economic calamities and by wars. The influenza epidemic of 1918 had heavy tolls of human life and many farms under the annual cropping had to be abandoned for some time (vide Dube, R.S. Population of the Rewa Plateau : A Geographical Analysis. Kanpur : Sahitya Ratnalaya, 1979). The historical evidences of retrograde progression of land use due to the destruction of population from war are available from the historical accounts of Malwa in Madhya Pradesh. The Malwa had a rich system of annual cropping which was completely destroyed by the forceful and trecherous Maratha invasion in the Medieval period and the land use system reverted back to the long fallowing.

CHAPTER FIVE

INNOVATIONS IN THE FARM TECHNOLOGY IN MADHYA PRADESH

Innovations in the farm technology are the essential prerequisite for increasing agricultural production. A full utilization of the farm potentials of a territory is not possible unless sufficient innovatory changes are introduced in agricultural techniques. Boserupean, the leading exponent of the dynamic theory of land use, has shown that in a pre-industrial society innovations in farm technology and the simultaneous changes in land use are stimulated by sustained increase in population density.[1] In the initial phases, when population is too sparse to cover all the cultivable parts of a territory for cropping, the long fallowing is practised, since this helps preservation of soil fertility without much input of labour and capital. The subsequent increase in population density, however, results in fallow shortening and a system of short-fallow cultivation fallows. The sustained growth of population eventually causes dimunition of a fallowing and a system of annual cropping to be followed later by multi-cropping comes into operation. These successive changes in the land-use systems are accompanied by significant changes in the farm technology in order to enhance intensity of agricultural production for supporting the growing population number.

The population of Madhya Pradesh has been registering rapid increase for a few decades. Its population has, infact, doubled within a fairly shorter period of thirty years from 1951 to 1981. The demographic dynamism seems to have caused significant changes in the land use and associated farm techniques. It would be interesting ot conduct a thorough evaluation of the farm technology in Madhya

Pradesh in order to verify the validity of Boserupean propositions and to highlight the need for quicker adoption of the new agricultural techniques lest the Malthusian system of population resource imbalances should overtake the Boserupean system of population resoure equilibrium.

Malthusianism v. Boserupeanism

The Malthusian pessimism on population resources imbalances is based on the idea that food production increases in the arithmetic progression, while population increases in the geometric progression leading eventually to a state of excess of people over the available resources.[3]

The views of Boserup are presented in sharp contrast with those of Malthus. Whereas Malthus states that food supply limits population size, Boserup propounds that population increase stimulates changes in agricultural techniques so that more food could be produced.[4] She examines the different land use systems of the world and suggests a close connection between the agricultural techniques and land use systems and states that any increase in the intensity of production and innovations in farm technology was unlikely, unless population increases. Her Book is based on the idea that techniques required by more intensive systems of land use are known to the people and as the population increases people are willing to adopt them. She cautious, however, that if the knowledge of the techniques is not available the agricultural system would regulate the population size—the Boserupean system would collapse to give way to the Malthusian forces.

Madhya Pradesh has been registering sustained increase in population for the last several decades. The new agricultural techniques were not available to the farmers of the state in the pre-independence period and Malthusian force had upper hand over the Boserupean forces to regulate the population size. In post-independence period the population pressure has succeeded in regulating the land use system with introduction of innovatory changes in farming as visualised by Boserup and the State registered sharp increase in per capita agricultural production from 164 kilograms in 1951 to 326 kilograms in 1961. It seems, however, that simultaneous growth in agricultural techniques has failed to occur to keep pace with the population growth after 1961 and the per capita farm output has declined to 295 kilograms in 1971 and 253 kilograms in 1981.[5] The sustenance of such a declining trend in population-food resource equilibrium is likely to result in the failure of the Boserupean system and the resurgence of the Malthusian system. The failure of swift adoption of technological innovations in Madhya Pradesh may look peculiar, since knowledge of the new techniques are avalable to the farming community, but it is reluctant and indifferent to adopt them with the consequence that labour productivity is gradually deteriorating. It would seem, therefore, to examine the temporal and spatial trend of technological changes on the farm in the state in order to identify the forces responsible for the failure of a quick transition in agricultural techniques and in order to verify the validity of the propositions advanced by Boserup.

Methodology

Reliable data on the successive changes in the farm technology in Madhya Pradesh are difficult to find. Conjectural conclusions may be made from the fragmentary archaeological and historical data, which may help surmising the passage of farm technology from the ancient digging stone stick-hoe phase of the long fallow system to the modern plough cultivation of the short-fallow cultivation systems. The paucity of data restricts our knowledge even for the modern periods and documentary evidences are not available for the period earlier than the year 1956-57. Any effort for the knowledge of the past farm technology would be, therefore, based on the present spatial ramifications of the different elements of farm techniques.[6] The available statistics on present mechanical and biological farm technology in Madhya Pradesh has, therefore, been used for the analysis of their spatial variations. The spatial variability has, then, been taken as a yardstick to identify the trend of spatial successions which suggest the trend of changes in farming techniques in Madhya Pradesh.

The technological innovations of farming have been classed into two groups,[7] the biological, which includes adoption of high-yielding varieties of crops, improved methods of crop fertilisation, and plant protection techniques, and the mechanical, which includes such new sources of mechanical power for farming as tractors, irrigation pumps and other more efficient iron-based farm implements. The use of biological innovations help increasing the farm productivity, while mechanical innovations save human labour and help in improving agricultural efficiency. The biological and mechanical techniques may, thus, help in improving farm efficiency and fetching greater farm output to support the rapidly growing population. The following sections of this paper highlight the present situation of adoption of technological innovations by the farmers of the state and include a discussion of the relevance of the theoretical propositions of the dynamic theory of land use and associated technological changes.

MECHANICAL TECHNIQUES

The principal farm implements being used in the farm in Madhya Pradesh include wooden ploughs, iron ploughs, tractors, bullock carts, irrigation pumps, expellers, plant protection machines and persian wheels, besides a number of older and smaller implements used for tilling and harvesting. These implements help in saving human labour and in enhancing the crop productivity. Some of the farm implements have been in use for many centuries, while some others have been introduced recently. An examination of the patterns of the individual farm implement in Madhya Pradesh may provide an understanding of the forces responsible for their innovatory changes and adoptions.

Wooden Ploughs

The greater part of the ploughing in Madhya Pradesh is accomplished from the traditional wooden ploughs. They are more numerous than any other ploughing equipment in the State (there are 232 wooden ploughs per 1,000 hectares of the cropped area). Their greater acceptance on the farm is attributable to a number of factors.

(*i*) The wooden ploughs are cheaper, and may be manufactured and repaired locally by the village artists.

(*ii*) A greater part of the farming needs shallow tilling of the farm soils. The wooden soil-eratchers are better suited for shallow tilling.

(*iii*) They are lighter to be drawn by the draught animals and for their transportation to the widely fragmented land holdings.

TABLE 5.1

Madhya Pradesh : Density of Mechanical Farm Implements Per 1,000 Hectares of the Cropped Area 1981

Machinery	Density per 1,000 hectares of the cropped area
Wooden ploughs	232
Iron ploughs	8.6
Tractors	1.7
Irrigation pumps	16.91
Bullock carts	111
Expellers	0.55
Plant protection machinery	1.07
Persian wheels	3.96

Source : Directorate of Agriculture, Madhya Pradesh, Agricultural Statistics, 1981.

(*iv*) They have been in use since antiquity and their substitution by modern ploughing equipments is difficult to achieve unless the whole socio-economic fabric of the conservative and traditional village community is fully altered.

The wooden ploughs are more numerous in the northern, eastern, and southern parts of the State (Map 5.1-A). The areas where their frequency per 1,000 hectares of the cropped area is higher than the State average of 232 include about 45 per cent of the total districts of the state. Their density is significantly smaller in the Lashkar region in the north and in the Malwa region and Narmada valley in the west. Their frequency is conspicuously high in the upper Wainganga basin, Baghelkhand plateau and Bastar plateau. The smaller density of the wooden ploughs in Lashkar, Malwa and Narmada region is attributable to a number of factors :

(*i*) These are relatively drier parts of the State and wooden soils eratchers are less suitable for the deeper soil tilling there,

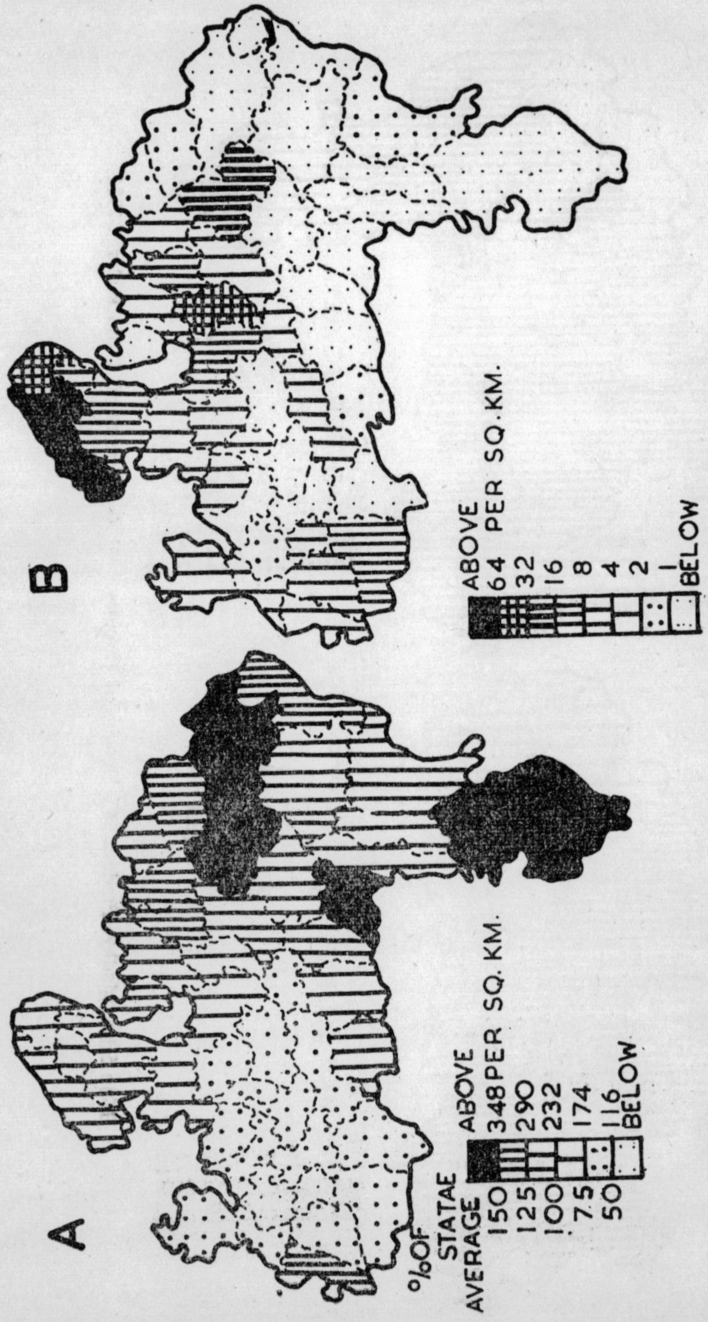

Map 5.1 : Madhya Pradesh A. Density of wooden ploughs, 1981 B. Density of iron ploughs, 1981.

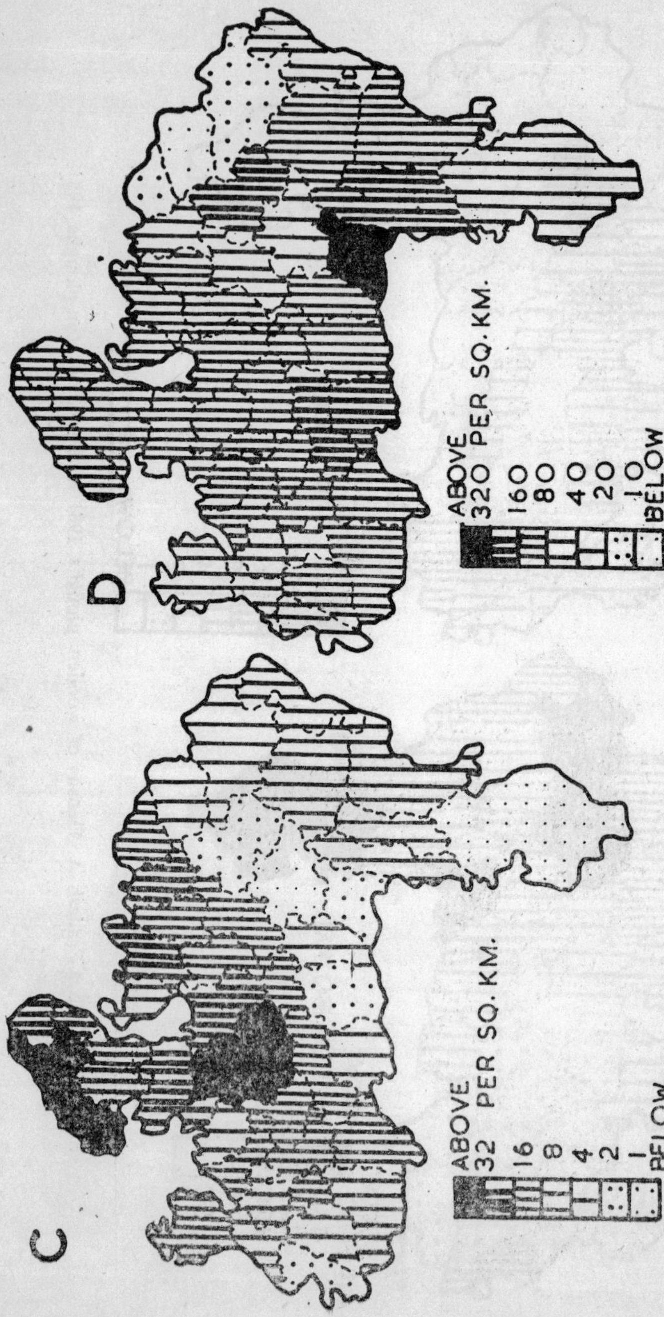

Map 5.1 : Madhya Pradesh C. Density of Tractors, 1981 D. Density of Bullock Carts, 1981

(*ii*) These are relatively less densely peopled, but more urbanized parts of the state. The human farm power is insufficient in these areas either due to the smaller population density or due to the diversion of rural labour to the industrial and urban activities, and

(*iii*) the size of the operational holdings in these areas is medium and large. The wooden ploughs are insufficient to cover the ploughing activity on relatively larger holdings.

A close observation of Map 5.1-A would suggest that density of wooden ploughs gradually increases from more developed areas with medium and large land holdings to the less developed areas of marginal and small operational holdings. The nature of terrain also seems influencing the spatial variation, for the hilly and dissected areas have greater density of ploughs than the areas of relatively homogeneous surface relief. This would suggest that the process of substitution of wooden ploughs has started from the more urbanized areas of the State and it is likely to expand to other parts of the State in future, though, forces of ignorance, conservation and socio-economic backwardness are in operation to restrict the growth of new farm technology.

TABLE 5.2

Madhya Pradesh : Number of districts under different classes of the Density of Mechanical Implements, 1981.

Density classes	*Wooden ploughs*		*Iron ploughs*		*Tractors*	
	class limits	*No. of dists.*	*Class limits*	*No. of dists.*	*Class limits*	*No. of dists.*
Very low	Below 116	2	Below 2	21	Below 2	5
Low	116-173	14	2-4	9	2-4	8
Medium	174-231	9	4-8	5	4-8	7
High	232-289	10	8-16	5	8-16	14
Very high	Above 290	10	16-32	1	16-32	6
Extremely high	—	—	above 32	4	Above 32	5

Density classes	Bullock carts		Irrigation pumps	
	Class limit.	No. of dists	Class limit	No. of dists.
Very low	Below 10	1	Below 2	1
Low	10-40	3	2-4	4
Medium	40-80	10	4-8	7
High	80-160	25	8-16	14
Very high	160-320	5	16-32	11
Extremely high	Above 320	1	Above 32	8

Source : Directorate of Agriculture, Madhya Pradesh, Agricultural Statistics, 1981.

The Iron made Farm Equipments

The iron made principal farm equipments include iron ploughs, tractors, irrigation pumps, bullock carts, expellers, plant protection machines and persian wheels. Density of farm implements, with exception of bullock carts is depressingly low in Madhya Pradesh. The per 1,000 hectares density of bullock carts is 111, while there are 17 irrigation pumps, 0.55 expellers, 1.07 plant protection machines, and 4 persian wheels per 1,000 hectares of the cropped area. Such a low frequency of modern farm implements suggests a very low degree of technological innovations in this State and much progress would be necessay if soils of State are to be subjected to their full utilization in the near future.

Though the data for spatial variations are available only for the iron ploughs, tractors and irrigation pumps, their regional contrasts are suggestive. The iron ploughs, tractors and irrigation pumps are more numerous in the areas lying north of the Narmada Son line. Their frequency is depressingly lower in the areas south of it (Map 5.1-B and C, 5.2C). The greater acceptance of the farm machinery seems to be associated largely with the size of operational holdins (Map 5.2-D) but in some cases population pressure has succeeded in bringing about replacement of older farm implements by the newer ones, irrespective of the size of land holdings. Two types of the areas seem to be discernible in this context :

(*i*) In the areas where land holdings are predominantly medium and large, and where high status farmers with their greater economic means predominate, the modern farm machinery has found greater acceptance than the areas where marginal and small land holding poor cultivators predominate. Malwa, Bundelkhand, Rewa and Narmada valley may be classed under this group.

(*ii*) In the areas where population pressure seems to be exceeding the resistance of marginal and small land holding farmers against the abundoning of conservative and traditional equipment, the older forms of farm machinery are being quickly replaced by the new farm implements. The northern parts of the Madhya Bharat plateau may be included under this class.

It may follow, therefore, that adoption of the innovations of the farm implements are succeeding more in the areas where size of land holdings is larger and where greater degree of urbanization and socio-economic development has made a speedier diffusion of the innovations.

BIOLOGICAL TECHNIQUES

The important biological techniques presently in operation on the farm in Madhya Pradesh include use of fertilizers, plant protection technology and application of high yielding seeds. These techniques are primarily yield-increasing and help in enhancing the farm productivity if they are supplemented by introduction of adequate water supply, mechanical techniques and improved agrarian structure.

Fertilizers

The farms of Madhya Pradesh consume about 238 thousand tonnes of chemical fertilizers (9.25 kilograms per hectare of the cropped area). The use of nitrogen is more favoured than that of the phosphate and potash. These nutrients are, in fact, used in the proportion of 8:4:1 respectively. The smaller use of fertilizers by the farmers in this state stems from a number of reasons. Firstly, the gross irrigated area covers barely 12 per cent of the gross cropped area. The chemical fertilizers may be gainfully used in the areas where irrigation facilities have been developed. Secondly,

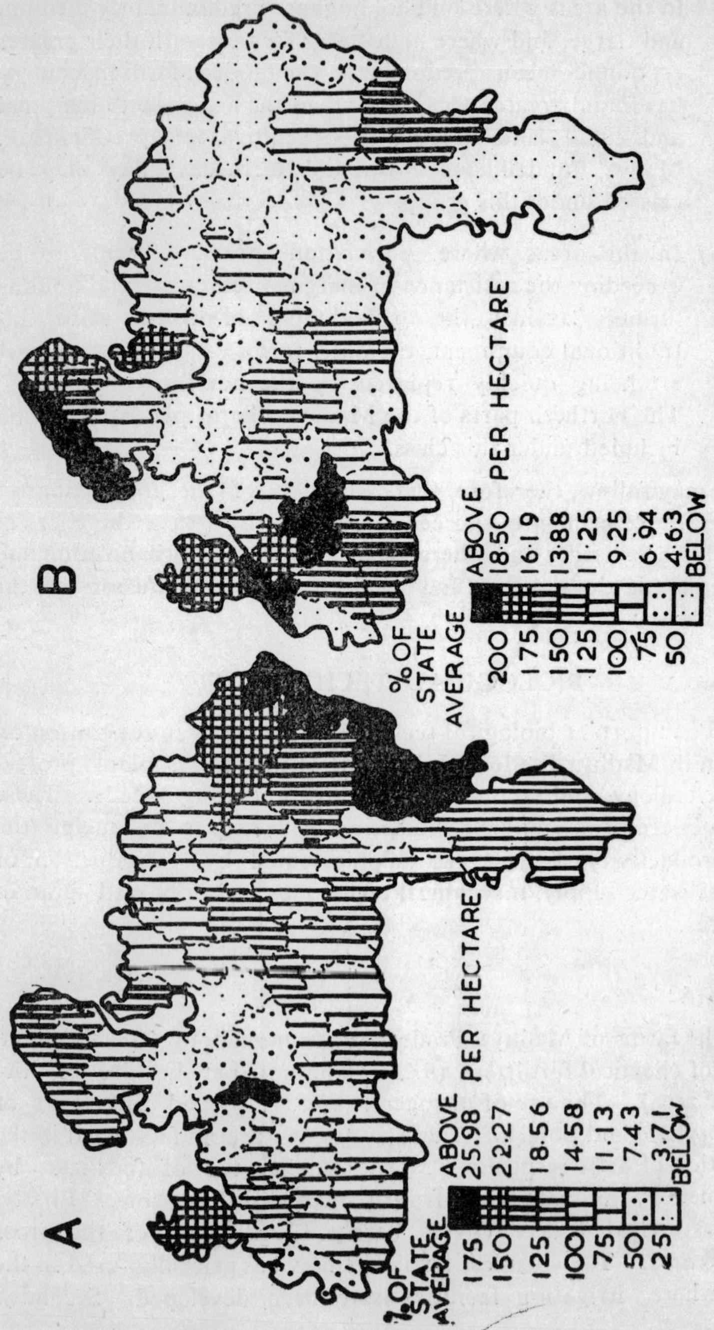

Map 5.2 : Madhya Pradesh A. High-yielded seed coverage, 1981 B. Use of fertilisers, 1981

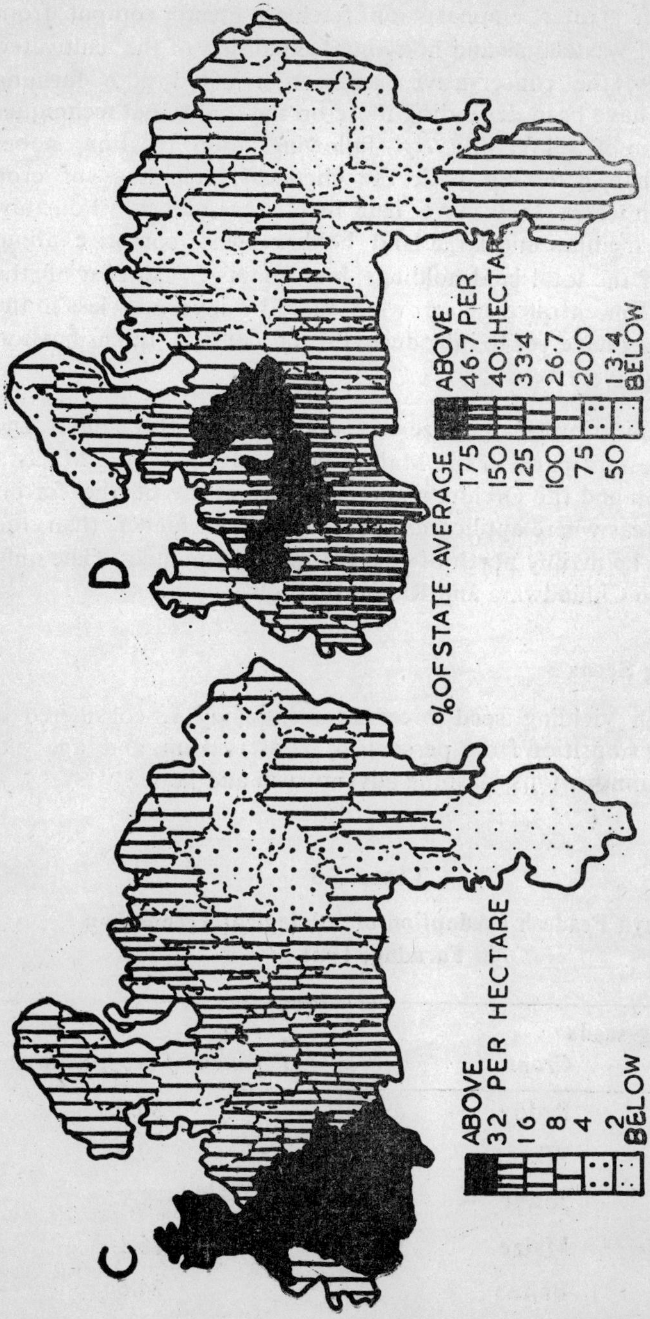

Map 5.2 Madhya Pradesh C. Density of irrigation pumps, 1981 D. Size of land holdings, 1981

there has been greater emphasis on fetching greater output from reclamation of wastelands and horizontal expansion of the cultivated area. Thirdly, the conservative, tradition oriented poor farming communities have been depending more on the traditional techniques of preservation of soil fertility, *e.g.* fallowing, crop rotation, gober etc. rather than gambling with the chemical techniques of crop fertilizers with which they have had little aquaintance. Fourthly, the efforts of medium and large land holders, who comprise about 27 per cent of the total land holdings but about 70 per cent of the total area, is concentrated on growing more by investing less in the forms and they have yet to introduce the agricultural intensification processes effectively.

The application of fertilizers is not uniform throughout the State. The western parts of the Malwa, northern parts of the Madhya Bharat plateau and the Orchha plain make greater use of the fertilizers. The areas where application of fertilizers is higher than the state average lie mainly north of the Narmada-Syon line. The only exceptions are Chhindwara and Raipur districts.

High Yielding Seeds

The high yielding seed breeding technology is considered a revolutionary transition from pessimism, conservatism and age old traditions to innovations, dashing adventurism and hope.[8]

TABLE 5.3

Madhya Pradesh : Adoption of Biological Technology in Farming, 1981

High yielding seeds		
	Crops	*Per cent of cropped area*
	Paddy	30.9
	Wheat	29.7
	Jowar	18.9
	Maize	16.1
	Bajra	24.6

Fertilizers	*Fertilizer*	*Kilogram per hectare of the cropped area*
	Nitrogen	5.72
	Phosphate	2.79
	Potash	0.74
	Total	9.25
Plant Protections		
	Items	*Per cent of total cropped area*
	Seed Treatment	5.68
	Crop Treatment	8.54
	Rat control	2.27
	Safe storage	11.05

Source : Directorate of Agriculture, Madhya Pradesh, Agricultural Statistics, 1981.

The introduction of high yielding varieties of crops cover about 31,22,000 hectares or about 15 per cent of the total cropped area in the State. The principal crops included under this programme are rice, wheat, jowar, maize and bajra. The improved varieties of rice cover about 15,03,000 hectares (31 per cent of the total area under the crop), followed by wheat (30 per cent), bajra (25 per cent), jowar (19 per cent) and maize (16 per cent).

The districts where the use of high yielding seeds cover greater part of the cropped area (more than the State average of 14.58) are more numerous in the area south of the Narmada-Son line, though some areas in the north also join the list of such districts e.g. northern part of the Madhya Bharat plateau, Orchha plain and a few districts of the Malwa. The greater adoption of high yielding seeds is associated with the areas where marginal and small land holdings are more. The high yielding seed techniques are more popular in the paddy areas. It appears that the high yielding seed technique is more acceptable to the low status poor farmers who can more profitably make use of the seeds in their small operational holdings without involving any substantial financial expenditure.

Plant Protection

The battle for plant protection is an important part of the high yielding variety. The plant protection programme includes application of seed treatment, crop treatment, wheat control, rat control and safe storage techniques. The seed treatment covered 12,15,000 hectares (5.7 per cent of the cropped area), the crop treatment 18,27,000 hectares (8.5 per cent), wheat control 107,000 hectares, rat control 486,000 hectares (2.3 per cent) and safe storage 23,65,000 hectares (11.1 per cent). The data related to the spatial aspects of the plant protection programme in different districts of the state are not available.

TABLE 5.4

Madhya Pradesh : Coverage of Biological Technology In Farming 1981 (As per cent of total cropped Area)

Classes	*High yielding seeds*		*Fertilizers*	
	Class limit	*No.of dists.*	*Class limit*	*No. of dists.*
Very low	Below 3.71	2	Below 4.63	8
Low	3.71-11.13	14	4.63-6.94	7
Medium	11.13-14.58	12	6.94-9.25	8
High	14.56-18.56	8	9.25-13.88	5
Very High	18.56-22.27	4	13.88-18.50	9
Extremely High	Above 22.27	5	Above 18.50	4

Source : Directorate of Agriculture, Madhya Pradesh, Agricultural Statistics 1981.

CONCLUSIONS

The preceding discussion on the innovations of farm technology in Madhya Pradesh highlights a depressingly low degree of the adoption of modern farm technology in Madhya Pradesh. The traditional technology is yet more acceptable to the low status cultivators with marginal and small land holdings and limited economic means which make them more tradition-oriented and poverty stricken. The innovations seen to be adoptable to the high status cultivators with usually a large farm and means to invest in farm operations. This difference is exemplified in the sharp regional contrasts in the adoption of farm innovations in the areas of marginal and small operational holdings at the one extreme and those of the medium and large holdings at the other extreme as discussed in the preceding sections of this paper. The innovations in the farm techniques in Madhya Pradesh are consequently occurring more as adaptation to the socio-economic status of the farming community rather than that of the degree of population concentration. The following remarks may seem relevant vis-a-vis the operation of Boserupean and Malthusian system of population resource equilibrium and technological innovations on the farm in Madhya Pradesh :

(i) The spatial variations in the farm technology in Madhya Pradesh do not correspond with the regional difference in population concentration. Some of the less peopled parts have registered greater adoption of the technological innovations (e.g. Malwa) than some of the more densely populated territories of the State (e.g. Chhattisgarh). The Boserupean sequence of change in the growth of farm technology would seem, therefore, failing in Madhya Pradesh since degree of population pressure does not seem causing simultaneous improvements in farm techniques.

(ii) The technological innovations in agriculture seem to be more closely associated with the size of land holdings and level of socio-economic development. The more urbanized western parts of the State with predominantly medium and large land holdings have registered greater innovatory changes in agricultural techniques than the relatively less urbanized eastern and southern parts where marginal and small holdings are more numerous. This may lead us to conclude that recent innovations have succeeded more in the areas where land holdings are relatively larger and where superior economic means of the predominantly high status farmers favour adoption of the costlier new farm techniques. A greater degree of innovations in some of the densely peopled areas with smaller land holdings e.g. Lashkar region would, however, suggest that population pressure subsequently overtakes the resistance and reluctance of the poorer and conservative farmers leading to the adoption of agricultural innovations irrespective of the degree of urbanization, size of land holding and level of socio-economic well being.

(iii) In the densely peopled areas where recent technological changes are failing to occur quickly, the Boserupean system of population resource equilibrium seem collapsing and the Malthusian forces resurging to regulate population size. The Chhattisgarh basin exemplifies the operation of such forces. The technological innovations and associated increase in productive intensity have lagged behind the rapid demographic dynamism there. The debilitating Boserupean system consequently seems leading the resurgence of Malthusian forces evidenced in the heavy depopulating movements of labour from there.[10]

(iv) The intermittent resurgence of the Boserupean and Malthusian forces in Madhya Pradesh seem following the sequence of course as outlined below :

(a) The Primitive phase, characterised by sluggish population growth regulated by Malthusian forces, since sufficient technological innovations are not available for increasing agricultural output. The examples of this phase are available in the hilly zone.

(b) The Transitional phase, when rapid population growth accelerates agricultural innovations. These innovations are introduced more quickly in the more urbanized areas and in the areas where high status mediun and large holding farmers are more numerous to adopt the costlier technology more promptly. The Malwa region gives an illustration of this phase.

(c) The Diminishing phase : The new agricultural techniques are not quickly adopted with the result that agricultural intensity fails to improve. The Boserupean system collapses. The Malthusian force supercede the Boserupean forces. Chhattisgarh exemplifies this phase.

(d) The Developing phase, when population pressure exceeds the resistance and reluctance of the conservative farming community and modern technology is eventually adopted to enhance agricultural productivity for supporting growing population. This would be the resurgence of the Boserupean forces overtaking the previous Malthusian forces. The Lashkar region in northern Madhya Pradesh illustrates the operation of this phase.

(v) It would follow from the preceding statements that the progression of the Boserupean population resource equilibrium in a region is frequently intercepted by the resurgence of the Malthusian forces. The history of population resource relations may therefore be viewed as struggle between the Boserupean and Malthusian forces which supercede one another intermittently during the long course of their historical development and growth.

References

1. Boserup, E. : Conditions of Agricultural Growth. Allen and Unwin, 1965.
2. Padmanabha, P. : Census of India, 1981, paper 2 of 1981 Provisional Population Totals.
3. Glass, D. V. (ed.) : Introduction to Malthus Watts, 1953.
4. Boserup, E. : *op. cit.*
5. Misra, R.P. : Population and Food Supply in Madhya Pradesh. Doctoral Dissertation, Saugar University, 1982 (unpublished).
6. Dube, R.S. : The Anomalous Growth of Fallowing in Madhya Pradesh, Geographical Outlook, Ranchi (forthcoming)
7. Singh, J. : An Agricultural Geography of Haryana, Kurukshetra Vishal Publications, 1976.
8. Singh, J. : *op. cit.*
9. Misra, R.P. : *op. cit.*
10. Misra, R.P. : Growth of Population in Madhya Pradesh, Uttar Bharat Bhoogol Patrika, Vol. 18, 1982, pp. 37-47.

CHAPTER SIX

THE CROPPING INTENSITY IN A BOSERUPEAN SETTING

Introduction

In a contrast to the Malthusian propositions, Boserup, the leading exponent of a dynamic theory of population-resource equilibrium, states that population increase stimulates changes in agricultural techniques and intensity of production, so that more food could be produced to support the growing population.[1] The essence of her ideas on population-resource equilibrium is based on the examination of different land use systems of the world, classified according to their intensity of production measured in terms of the frequency of cropping.[2] At the one extreme of the land use systems are pointed out the forest fallows associated with shifting cultivation of least productive intensity with the multiple cropping systems which yield more than one harvest a year arranged at the other extreme. There is close connection between the agricultural techniques and the land use system.[3] The innovations in the agricultural techniques and resultant increase in agricultural productivity is suggested to occur under the impact of forces emanating from the sustained increase in population.[4] In fact, as population increases beyond a certain point, where no extra land is available for agricultural expansion, the length of fallows is shortened to maintain the previous level of per capita consumption. The diminution of the fallows and continuous use of the cultivated land for cropping causes decline in soil fertility and output per man hour.[5] The adoption of new techniques, however, becomes necessary and agricultural development and growth of food supply fallows simultaneously with the growth of population in a territory.

Madhya Pradesh affords an excellent example of a developing society where Boserupean forces are currently in operation.[6] It would be interesting to examine the cropping intensity in Madhya Pradesh in all of its aspect with a view to substantiate the validity of the Boserupean postulates and associated forces of population resource equilibrium. It would help highlighting the need for a speedier intensification of agricultural production best resurgence of Malthusian forces should overcast the Boserupean population resource belances.

The Boserupean Setting in Madhya Pradesh

The Boserupean forces are presently in operation in Madhya Pradesh.[7] This is suggested by a close scrutiny of the demographic and agricultural setting of the state.

Demographic setting

Demographic data for the pre-census period are not available for Madhya Pradesh. Indirect sources, however, suggest a sluggish but sustained increase in population during the whole extent of the historical period.[8] Though, the history of population growth in this part of the country has been frequently marked by the occurrence of such depopulating forces as draughts, famines, epidemics and political unrests, the overall trend during a longer period of time has been that of a sustained increase in population number.[9] The growth of population during the modern census taking history of the state is suggestive and needs a brief mention here. The population has increased more than three-fold during the period 1901-1981 registering a down trend only in 1921 when population decreased due to the widespread outbreak of devastating influenza, epidemic in 1918 (Table 6.1).

TABLE 6.1

Madhya Pradesh : Trend of Population Growth 1901-1981

Year	*Population (millions)*	*Density (per kilometre of the cultivable area*
1901	16.86	65.8
1911	19.44	75.9
1921	19.17	74.8
1931	21.36	83.4
1941	24.00	93.6
1951	26.07	101.8
1961	32.37	126.4
1971	41.65	162.6
1981	52.13	203.5

Source : Calculated from Census and Landuse data 1901-1981.

The population has had a continuous uptrend since then and its rapid growth has been causing strains on the agricultural resources. The pressure on per square kilometre of the cultivable area has infact, increased from 65.8 persons in 1901 to 203.5 persons in 1981. Such an increase in the pressure of population would exert significant impact on the land use system.

The population is unevently distributed in Madhya Pradesh. The density map of the state prepared on the basis of the composit scope of the Kendall's ranks of the district in respect of arithmetic density, rural density, agricultural density, and nutritional density (Map. 6.1A) suggests that High density areas of the state are associated mainly with the northern part of the Madhya Bharat plateau, Orchha plain, Rewa plateau, parts of Narmada Valley, Upper Waingarga Basin and Chhattisgarh Basin. The Low density belt forms a continuous stretch from the Madhya Bharat plateau and Bundelkhand upland in the north to the Satpura-Maikal ranges in the south. The Bastar plateau is also closed under the Low category. It would be interesting to examine the *inter se* spatial correspondence between the density of population on the one hand and intensity of cropping on the other in the effort to identify existence of Boserupean forces of population resource relation.

The Land Use System

Fragmentary archaeological and historical evidences suggest a change in the agrarian system of the state ever since the introduction of cropping in the neolithic period.[11] The information show that land use systems have registered significant changes from the old forest fallowing phase to the present manual cropping phase through the intervening phases of fallowing and short-fallow cultivation systems.[12] The availability of data on recent changes in land use are restricted for the period 1956-1981, but the trend of changes revealed by them is suggestive to the existence of the operation of Boserupean sequence of population resource changes.

It may be seen from the Table 6.2 that fallows (9.46 per cent) and wastes (16.73 per cent) occupied quite a substantial part of the cultivable area of the state in 1956-57. Their share has gradually declined in 1980-81 suggesting that simultaneous increase in population necessitates shortening of the wastes and the fallows to permit horizontal expansion of the cultivated area for fetching more agricultutal production. Correspondingly, the net sown area has registered significant improvement from 62.19 per cent in 1956-57 to 73.45 per cent in 1980-81.[13]

TABLE 6.2

Madhya Pradesh : Growth of fallows and Wastelands 1956-1981
(Percentage of the cultivable area)

Year	*Cultivable land*	*Fallows*	*Cultivable wastes*	*Net sown area*
1956-57		9.46	16.73	62.19
1957-58		10.73	15.06	60.43
1958-59		9.44	14.39	69.91
1959-60		9.53	13.45	63.11
1960-61		11.03	12.30	63.60
1961-62		8.13	11.35	64.86
1962-63		7.42	9.77	60.23
1963-64		7.83	9.74	66.01
1964-65		7.06	9.82	66.43
1965-66		7.97	9.27	66.45
1966-67		8.25	9.22	68.29
1967-68		7.01	9.10	70.52
1968-69		6.43	9.04	71.54
1969-70		6.09	8.89	72.65
1970-71		6.22	8.34	72.66
1971-72		6.12	8.12	73.07
1972-73		6.46	8.14	72.97
1973-74		6.50	8.35	73.02
1974-75		6.78	7.80	72.90
1975-76		6.40	7.94	73.56
1976-77		7.33	8.11	72.94
1977-78		6.93	7.54	74.19
1978-79		7.06	7.29	74.30
1979-80		8.76	7.53	72.40
1980-81		7.92	7.44	73.45

Source : Directorate of Agriculture, Madhya Pradesh, Agricultural Statistics, 1981.

The changes in the farm techniques also suggest the continuation of Boserupean dynamism in Madhya Pradesh.[14] Though information on the previous technological changes are difficult to find out the present spatial ramifications of technological innovations suggest a complete sequence of changes in agricultural techniques. It has been observed[15] that the present farm operation are predominantly based on the plough cultivation system, with the use of such old implements as digging sticks and hoe in some of the tribal and hilly enclaves of the state. The present innovations[16] bringing out transition from the ploughing phase to the tractorisation and intensification phase also manifest the change in agricultural techniques necessitated by increasing pressure of population on the farm resources.

It may, therefore, be concluded that Madhya Pradesh is currently evidencing a transition in the Boserupean land use system and analysis of the cropping intensity may bring out such facts on the population resource relations which may illuminate the changing relations between population and land use systems.

Agricultural Expansion

Expansion of the cultivated area is an important measure for enhancing agricultural intensity in the early phases of the Boserupean setting of the population resource equilibrium. We do not have precise and reliable data on the ancient forest and bush fallowing phase in Madhya Pradesh. Conjectural conclusions may help in concluding that frequency of cropping was successively increasing with the growth of population in different parts of the state and also that the rapid shortening of the fallows was occurring with the acceleration of population. Expansion of the cultivated area for creation of new fields at the so-called extensive margins[17] of the crop land for accommodating the growing pressure of population has been evidenced during the period 1956-1981 for which reliable land use statistics are available. In the year 1956-57 cultivated area comprised about 71.65 per cent of the total cultivable area of with an extensive scope for expansion of agriculture on about 75.37 thousand hectares of the wastes and other uncultivable lands (or about 29 per cent of the cultivable area). The cultivated area has registered significance increase since than (Table 6.2).

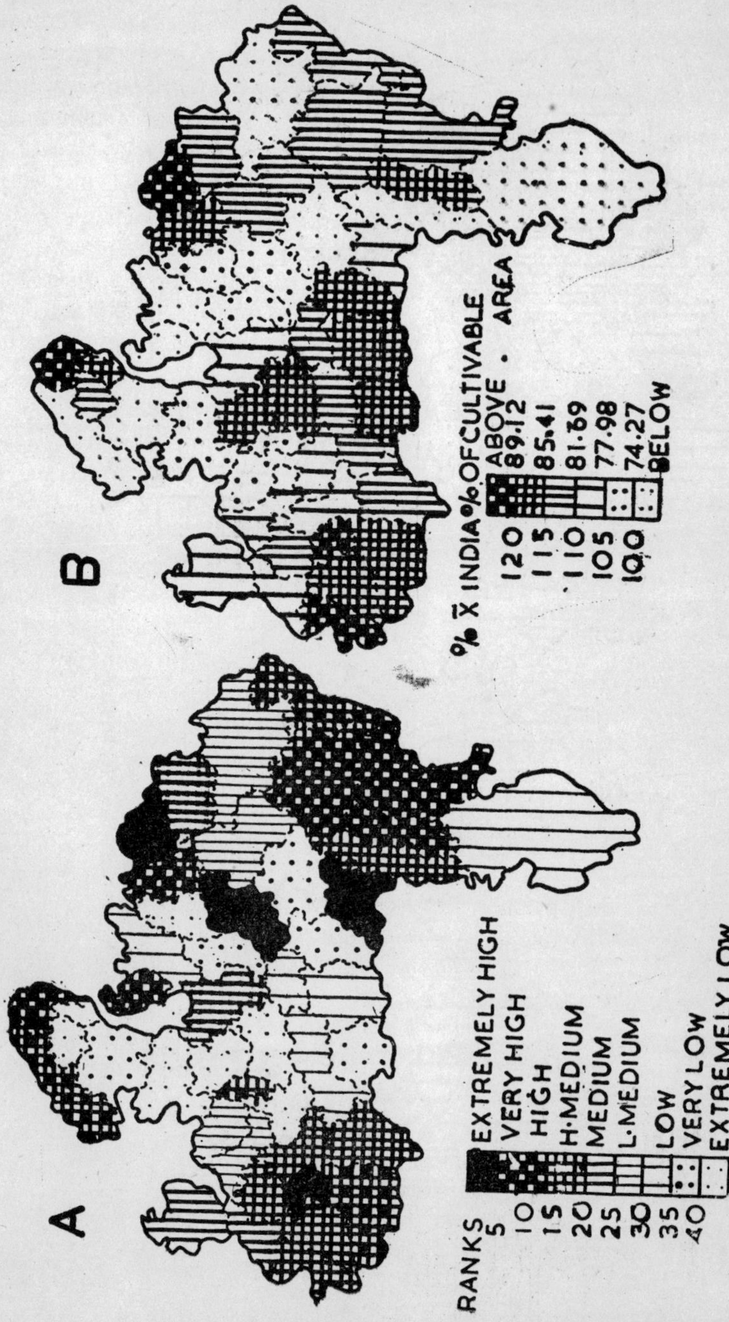

Map 6.1: Madhya Pradesh A. Population Density Regions, 1981 B. Cultivated Area, 1981

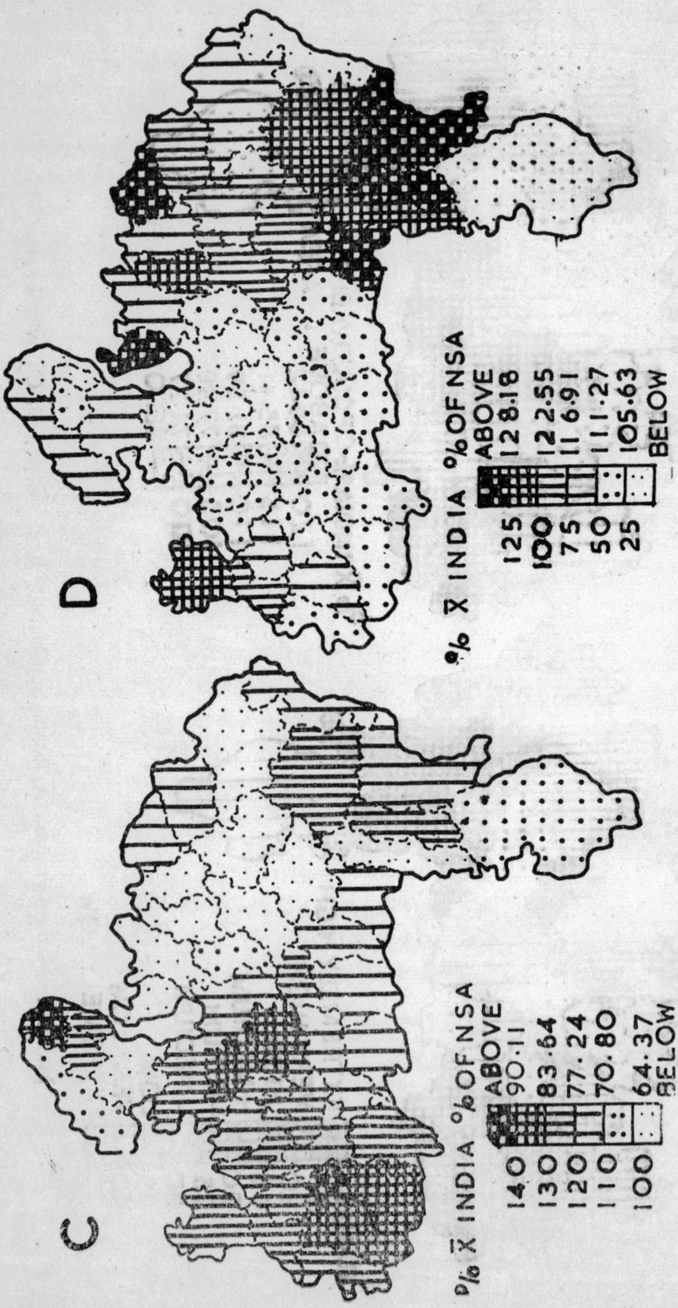

Map. 6.1 : Madhya Pradesh C. Net sown Area, 1981 D. Cropping intensity, 1981

With the result that it now occupies about 81.36 per cent of the total cultivable area. The culturable wastes have registered considerable shortening. They have been reduced to 7.44 per cent of the cultivable area in 1981 from 16.73 per cent in 1956-57. Such a rapid reduction in the wastes and simultaneous increase in the cultivated area would suggest an occurrence of rapid agricultural expansions as an important measure of increasing agricultural production to support the stupendously growing population of the state.

An observation of Map 6.1-B would suggest that share of cultivated area is higher than the national average in 37 districts of the state. The proportion of cultivated area ranges from 62.7 per cent in Shivpuri district to 90.7 per cent in Bhind district suggesting a wide range of spatial variation of the operation of Boserupean forces responsible for agricultural expansion in this state. The areas where share of cultivated area is relatively higher (more than 85.41 per cent) include Lashkar plain, Rewa plateau, eastern fringes of the Malwa Western margins of the state, Satpura region and some part of the Chhattisgarh region. There is no close connection between the degree of population concentration and degree of agricultural expansion in different parts of the state, the densely peopled Chhattisgarh region (Map 6.1-A) has registered a relatively lower incidence of expansion of the cultivated (Map 1-B) and some sparsely peopled areas (Satpura region) have registered higher incidence of agricultural expansion as a measure of increasing agricultural intensity. Existence of such a spatial anomaly between population concentration and agricultural growth may suggest that the Boserupean postulates are dubious in respect of *inter se* relation between agricultural expansion and population growth in Madhya Pradesh.

Changes in net Sown Area and Fallowing

In consonance with the occurrence of agricultural expansion under the sustained increase of population pressure, the net area sown has registered rapid improvement and fallowing has evidenced substantial reduction in Madhya Pradesh during the period 1956-1981 (Table 6.2). With minor fluctuations, the share of net sown area in the cultivable area has increased from 62.19 per cent in 1956-57 to 73.45 per cent in 1980-81 registering a change of about 120.50 percent The increase in the share of net sown area during the period 1956-81 was, however, greater than that of the cultivated area, suggesting reduction of the fallowing activity. The fallows have been reduced to 8.92 per cent in 1980-81 from about 9.46 per cent in 1956-57. The expansion of annual cropping and reduction of fallowing, may result in the rapid exhaustion of soil fertility in the absence of a quick improvement in agricultural techniques. In fact, such a situation seems developing in Madhya Pradesh for some time causing retrogration of the fallowing activity.[19] The share of fallows which had a declining trend upto the year 1972 started increasing subsequently (Table 6.2) perhaps due to the exhaustion of soil fertility or depletion of farm labour.[20] Such an inverse trend of fallowing is peculiar and has remained unknown and unreported in geographical literature. This may suggest a debilitating operation of Boserupean forces gives way to the Malthusian forces in Madhya Pradesh since a retrograde growth of fallowing is difficult to envisage in the Boserupean setting. A quick adoption of the process of agricultural intensification would be necessary to avert the collapse of Boserupean forces in Madhya Pradesh.

Over a greater part of the state the share of net sown area in the cultivable area is greater than the national average of 64.37 per cent (Map 6.1-C). The share of net sown area is high (more than 83.64 per cent of the total cultivable area) in the parts of Malwa and adjoining tracts. It is relatively lower in other parts of the state. There is again no evident correspondence between percentage of the net sown area (Map 6.1-C) and degree of population concentration (Map 6.1-A). In fact many more densely peopled areas e.g. Lashkar, Narmada, Chhattisgarh and Rewa regions support a relatively smaller share of the net sown area. This again appears disproving the Boserupean propositions in Madhya Pradesh.

AGRICULTURAL INTENSIFICATION

Intensification of cultivation or the land use efficiency which measure the extent to which the net sown area is cropped,[21] is one of the essential pre-requisites for the sustenances of the Boserupean population resource equilibrium. Failure in agricultural intensification is bound to cause collapse of the Boserupean system to give way to the resource of Malthusian forces to endanger the survival of mankind.

There is only modest scope now for expansion of cultivation by creation of new farms at the outer margins of the existing croplands of the state. The cultivable wastes now occupy barely 7.44 per cent of the total cultivable area of the state (Table 6.2) and the increase of population number in future would necessitate intensification of cropping rather than expansion of cultivation. Presently multicrop ping is restricted to only 74.43 per cent of the total net sown area of the state. An observation of Table 6.3 would suggest that there has been considerable temporal interruptions in the growth of multicropping and there has been little change in the share of multicropped area in 1980-81 as compared with that in 1956-57. Multicropping included about 13.61 per cent of the net sown area in 1956-57. After evidencing considerable fluctuations it now occupies about 14.43 per cent of the net sown area in 1980-81. This would suggest that : (*i*) the multicropping activity has not yet stabilised even in the areas where it is practised, (*ii*) the multicropping activity has yet to expand to overtake the agricultural system from the annual cropping system, and (*iii*) absence of swift transition from annual cropping to multicropping, under a state as sustained population growth as being evidenced in Madhya Pradesh may disprove the Boserupean contentions on intensity of cropping.

The areas of higher cropping intensity in Madhya Pradesh are found in the Chhattisgarh basin, Rewa plateau, Orchha plain and Mandsaur region. A comparison of the Map 6.1-D with 6.1A would suggest absence of any close relation between degree of population concentration and cropping intensity, since some of the densely peopled parts e.g. Lashkar plain in the north, have registered very low cropping intensity. The spatial variations of the cropping intensity seem giving increase correlation with the size of land holdings. The spatial ramifications of the cropping intensity also seem failing to fully support the Boserupean propositions.[22]

Agricultural Productivity

There has been net on average in agricultural output from about 90,59,000 tonnes of food crops in 1956-57 to 13,020,000 tonnes in 1980-81 in consonance with the uptrend of demographic growth. The per hectare price weighted agricultural productivity, however, shows no increase during the period 1956-81 suggesting that emphasis of farming in this period has been on raising agricultural output by expansion of cultivated area rather than from agricultural intensification. The trend of price weighted per hectare

TABLE 6.3

Madhya Pradesh : Growth of Multicropping 1956-1981
(Percentage of NSA)

Year	*Multicropped area %*
1956-57	13.61
1957-58	8.78
1958-59	12.95
1959-60	13.59
1960-61	12.97
1961-62	13.87
1962-63	12.55
1963-64	13.44
1964-65	13.31
1965-66	7.60
1966-67	6.57
1967-68	10.31
1968-69	10.61
1969-70	11.11
1970-71	12.04
1971-72	13.15
1972-73	12.14
1973-74	14.28
1974-75	10.77
1975-76	14.11
1976-77	12.53
1977-78	14.41
1978-79	15.38
1979-80	13.18
1980-81	14.43

Source : Directorate of Agriculture, Madhya Pradesh, Agricultural Statistics, 1981.

agricultural output during the period 1956-81 measured at the static price rates of 1980-81 are summarised in Table 6.4. For the evaluation of price weighted agricultural productivity, 18 crops have been considered. These are rice, jowar, bajra, barley, wheat, maize, gram, tur, urd, moong, masoor, groundnut, sesamum, rape and mustard, linseed, castor, cotton and sugarcane. The important features seem emerging from the data recorded in this table :

(i) the agricultural productivity has had a serrated growth characterised by frequent rise and fall in the growth curve. This has been caused, perhaps, because of fluctuations in the weather conditions and because of frequent changes in the composition of crops produced; and

(ii) there has been no change in the price weighted per hectare productivity of crops was on 1148.50 in 1956-57 which has registered very little increase in 1980-81 when it was Rs. 1183.60. Such a temporal change would suggest, therefore, absence of any demographic relevance with the growth of agricultural productivity.

Similar conclusions also emerge as re-spatial ramifications of agricultural productivity, measured in terms of Singh's crop yield and concentration indices coefficient,[23] show sharp regional variations (Map 6.2). The method involves the calculation of crop yield index (formula $Yi = \frac{Ya_c}{Ya_r} \times 100$, where yi is the crop yield Index, Ya is the average yield of 'a' in the component enumeration unit and Ya_r is the average yield of the crop 'a' in the entire state and crop concentration index (formula $C_1 = \frac{Pa_c}{Pa_r} \times 100$, where C_1 is the crop concentration index, Pa_c is the percentage strength of the crop 'a' in the total harvested area in the enumeration unit, and Pa_r is the percentage strength of the crop 'a' in the entire state for all the individual crops of the districts. Rank of a district in respect of crop yield index and crop concentration index of a crop have been then summed and averaged to give the crop yield and concentration indices Ranking coefficient for a crop. The Ranking coefficient of all the crops have been summed and averaged to give a composit ranking coefficient for districts. The districts then ranked and mapped to give the regional pattern shown in Map 8.2. Cereals, pulses and oil-seed have been involved in the determination of composit Ranking coefficient for different districts of the state.

A careful look on the Map 8.2 would reveal that the areas of Higher crop production extend in parts of Malwa, Bundelkhand and Madhya Bharat plateau, essentially north of the Narmada. Rest of the state i.e., east of the Ken and south of the Narmada is classed under different classes of Medium and Low categories adopted for regionalisation of the Ranking coefficients.

A comparative study of this map to Map 6.1-A would suggest absence of any spatial correspondence of agricultural productivity with degree of population concentration, since such densely populated regions as Rewa plateau, Upper Wainganga plain and Chhattisgarh basin are classed under different categories ranging from conspicuously low to low medium productivity and relatively less density area of Malwa and southern Madhya Bharat plateau are classed as high to conspicuously high classes of agricultural productivity.

TABLE 6.4

Madhya Pradesh : Trend of Price Weighted Agricultural Output 1956-81 (at the static rate of 1980-81)

Year	*Per hectare output in Rs.*
1956-57	1148.5
1957-58	953.1
1958-59	1259.2
1959-60	1200.4
1960-61	1155.3
1961-62	1184.8

1962-63	1077.9
1963-64	1109.5
1964-65	1189.3
1965-66	925.2
1966-67	779.6
1967-68	1005.2
1968-69	969.3
1969-70	1070.3
1970-71	1156.8
1971-72	1199.1
1972-73	1090.2
1973-74	1114.4
1974-75	923.3
1975-76	1299.5
1976-77	1017.0
1977-78	1224.0
1978-79	1151.5
1979-80	803.7
1980-81	1183.6

Source : Directorate of Agriculture, Madhya Pradesh, Agricultural Statistics, 1981.

Conclusions

The preceding discussion on the Boserupean setting studied in respect of temporal and spatial ramifications of the intensity of production lead to the following conclusions :

1. The increase of population seems affecting the temporal trends in agricultural expansion in Madhya Pradesh, since significant horizontal expansion of the cultivated area has occurred during the period 1956-1981. Temporal trends confirm Boserupean but spatial pattern disproves corres-

pondingly, significant increase in the net sown area and reduction in fallowing and waste lands has also taken place. The spatial patterns of the share of cultivated and net sown area, however, no spatial correspondence with the degree of population concentration.

2. The agricultural intensification measured in terms of the cropping intensity has not increased in parallel succession with the growth of population during the period 1956-81. The state has not yet made sufficient changes from the present annual cropping system to multi-cropping phase. The spatial variation, of the cropping intensity also fails to show its close relation with spatial patterns of the degree of population concentration.

3. The emphasis of farming in Madhya Pradesh during the period 1956-1981 has been mainly on increasing agricultural output by agricultural expansion of the cultivated area and there has been little change in the level of agricultural intensification as suggested from absence of any substantial increase in the share of multi-cropping area and price-weighted per hectare agricultural output during the period 1956-1981. The spatial variations of price-weighted productivity also have no close relation with spatial changes of population density.

4. The failure of the agricultural intensification to increase in parallel succession with population growth during the period 1956-1981 may seem to be exerting operation of a debilitating influence on the operation of Boserupean system of population resource equilibrium in Madhya Pradesh. The agricultural planners may be advised to take note of the trend since this may accelerate the Malthusian forces to operate in regulation population size.

5. The agricultural conditions in Madhya Pradesh seem to suggest that there has been significant increase in the share of cultivated area and net sown area with simultaneous reduction in the wastes and fallows during the period 1956-1981. The agricultural output has consequently increased confirming the validity of Boserupean propositions at the

temporal scale. Their spatial variations, however, show little *inter se* correspondence with the degree of population concentration exemplifying failure of Boserupean postulate in the regional context of Madhya Pradesh.

References

1. Boserup, E. : Conditions of Agricultural Growth. London : Allen and Unwin, 1965, pp. 11-14.
2. *Ibid.* pp. 15-16.
3. *Ibid.* pp. 26-27.
4. *Ibid.* pp. 65-69.
5. *Ibid.* p. 117.
6. Dube, R.S. : Innovations in the Farm Technology in Madhya Pradesh : Uttar Bharat Bhoogol Patrika (forthcoming).
7. *Ibid.* and Dube, R.S. : "The Anomalous Growth of fallowing in Madhya Pradesh". Geog. Out. (Forthcoming).
8. Davis, K. : The Population of India and Pakistan. Princeton University Press, 1941.
9. Dube, R.S. : 'Population of the Rewa Plateau (Rampur : Sahitya Ratnalaya, 1979).
10. Kendall.
11. Dhavalikar, M.K. & Gregory, L.P. : 'Subsistence Pattern of an Early Farming Community of Western India.' Puratattva No. 7, 1974.
12. Dube, R.S. : "The Anomalous Growth of Fallowing in Madhya Pradesh". Geog. Out. (Forthcoming)
13. Dube, R.S. *op. cit.*
14. Dube, R.S. *op. cit.*
15. Dube, R.S. *op. cit.*
16. Dube, R.S. *op. cit.*
17. Boserup, E. *op. cit.* p. 12.
18. The Regional Nomenclature adopted in this paper are based on the physical plates of the National Atlas Organisation Calcutta.
19. Dube, R.S. *op. cit.*

20. Dube, R.S. *op. cit.*

21. Sing, J. : An Agricultural Geography of Haryana. Kurukshetra Vishal Pub., 1976, p. 235.

22. Dube, R.S. *op. cit.*

23. Dube, R.S. *op. cit.*

24. Sing. J. : An Agricultural Geog. of Haryana (Kurukshetra Vishal Publishers 1976) pp. 318-319.

Chapter Seven

THE MEASUREMENT OF AGRICULTURAL PRODUCTIVITY

The agricultural productivity is a qualitative term which provides an estimate of the power of agriculture to produce crops. It is not a synonym of soil fertility which expresses the ability of soils to provide plant nutrients. The notion of productivity refers to the efficiency with which inputs are utilised in agricultural production (Shafi, 1967). It measures the ratio of the index of agricultural output to that of total inputs in the farm production and therefore, provides an estimate of the farm output per unit of input (Pandit, 1965). Saxon (1965) is of the view that productivity is a physical relationship between output and the input which gives rise to that output.

The measures, of agricultural productivity which have been frequently adopted in different regional studies of agriculture are usually those of 'partial productivity' (Shafi, 1971), since only a single input or a group of inputs have been taken into account for the determination of farm productivity. Herring (1964) has categorized the farm inputs into three main classes—the land, labour and capital. A composit measure, where all these components of agriculture could be involved are difficult to be adopted perhaps for two reasons. Firstly, the data required for the measurement of productivity, a single input is more likely to be available than those required for the evaluation of overall productivity. Secondly, the aggregation of total inputs may tend to obscure the effects of changes in their composition.

Among the various inputs, land is the most permanently fixed and has assumed special importance in the regional studies of agriculture. It would not look peculiar that many of the measures of productivity are based on the land component of the farm inputs. In the absence of any measure for the evaluation of overall productivity, it must be appreciated that the choice of productivity measure would depend mainly, on the purpose for which its estimation is desired. The land factor would serve the purpose better, for instances where the land reforms and improvements in farm technology are to be affected. The labour productivity could be applied where knowledge of the income of population engaged in agriculture is desired. The cost and benefit situation of farming may be obtained better from the consideration of the capital component of agricultural inputs. It may be noted, however, that capital investment structure is often so much complicated that capital based productivity is difficult to compute and interpret.

The problem of the choice of measures of agricultural productivity was dicussed at great length at the 23rd annual conference of Agricultural Economists and it was generally agreed that land factor may be recognised as the principal basis for evaluation of agricultural productivity (I.S.A.E. 1964). The yield rate may, therefore, be considered a good measure of productivity while other factors of production may be taken into account while making regional comparison of farm production.

Approaches to the Measurement of Productivity

Efforts for identifying an aggregate measure of agricultural productivity are seen in a number of agricultural studies. Thompson (1926) has measured it in terms of gross output of crops and livestock by making use of the following parameters :

(*i*) per acre yield of crops;

(*ii*) density of livestock per 100 acres;

(*iii*) gross production *i.e.* output per 100 acres of the farmland;

(*iv*) percentage of arable land;

(*v*) number of persons employed in agriculture;

(*vi*) cost of production in respect of wages, labour-cost, rent or interest ; and

(*vii*) prices, relative profitability and general economic conditions.

Kendall (1979) initiated a system of four coefficients—a productivity coefficient, a ranking coefficient, money value coefficient, and a starch equivalent, *i.e.* energy coefficient. He took the acre yields of crops and ascertained their rank positions which were averaged to give ranking coefficient of agricultural efficiency. Ganguly (1938) has tried to measure the productivity by considering the area and yield of crops. His measure is based on the computation of the area under a particular crop as a percentage of the area under all the selected crops, and determination of the index number yield found out by dividing the yield of a crop in a unit area by the yield of the same crop in the entire region. The productivity was obtained by multiplying the proportion of area under a crop and the corresponding index number of its yield.

The agricultural efficiency of Uttar Pradesh was later determined by Shafi (1960) on the basis of the ranking coefficients of the acre yields of the selected crops. Enyedi (1964) while discussing geographical types of agriculture, prefers the use of a formula for determining an index of productivity.

Coefficient : $\frac{Y}{Yn} \frac{T}{Tn}$, where Y=The total yield of the respective crop in the unit area ; Yn=the total crop on the national scale ; T=Total crop area of the district ; and Tn=Total crop area on the national scale. Shafi, 1971 has later tried to analyse the regional dimensions of agricultural productivity in India by making use of Enyedi's Index of productivity.

Hirch (1943) has suggested the use of a crop yield Index determined as the average of the yields of various crops on a farm in an area relative to the yields of the same crops on another farm of the same area or another area.

Zobel (1950) prefers the use of labour productivity in the farm of output per man hour. He determines it as the ratio of total output to the total man hour envolved in agricultural production expressed as follows :

$\pi = f$ (P.L.), where f=Labour productivity

P=Production

L=Labour utilised in men hours.

Dondyal (1964) has measured variations in agricultural development and productivity taking into account the consideration of the role of credit, intensive crop enterprise and the influence of irrigation. Yang (1965), however, makes a substantial contribution to the measurement of farm productivity. He calculates index on the average yield basis of different crops selected for enumeration on a unit area and the yields of those crops in the entire study area. This is done by dividing the average yield of the crop in the entire study region. This is multiplied by 100 to the index number as shown in column 5 of the Table 7.1. The production is obtained by taking the area under selected crops as weight and multiplying it with the index number. The final crop index is obtained by summing the products of different crops and dividing it by total area under all the crops.

TABLE 7.1

Method of Calculating Crop Yield Index of X Farm

Crop	*Yield per hect. (in quintals)*		*Area under the crop on Farm X*	*Crop yield on farm X as % of the region (Col 3 ÷ col. 2×100)*	*Percentage multiplied by area under crop (Col. 4×5)*
	Entire Region	*Farm X*			
1	2	3	4	5	6
Rice	27	25.5	15	94	1410
Jowar	33	30	7.5	91	682.5
Wheat	30	33	15	110	1650
Gram	45	54	22.5	120	2700
Total	—	—	60	—	6442.5

Yield Index=6442÷ 60=107.4%

Stamp (1958) has preferred the use of standard nutritional unit, where per hectare food production is converted into calories and on the basis of minimum colorie requirement of an average person the average number of persons who can depend on the food produced in the unit area are considered. This is termed as SNU.

THE PROBLEM OF FINDING A COMPOSIT MEASURE OF PRODUCTIVITY

The International Commission on Agricultural Typology is making efforts for finding an aggregated measure of agricultural productivity. Kostrowicki suggests the use of three groups of criteria for evaluation of agricultural typology, viz. Social and land characteristics ; Organisational-cum-technical characteristics and Production characteristics. The principal problem in case of productivity estimates is not that of choice of factors, but it is what made use of their aggregation into a productivity index. Some workers have of Kendall's composit scores for regionalisation of the level of agricultural productivity. Dube (1984), for instance, has evaluated agricultural productivity in different districts of Madhya Pradesh by making use of Yang's crop yield Index, Stamp's SNU and price weighted farm output (in terms of output per hectare and per agricultural worker). The districts were ranked separately for each measure and their ranking scores for each district were totalled and averaged to give a composit score for each district. These composit scores have been mapped by the author to highlight areal variations in different parts of the State.

The use of population potential model (Stewart, 1958) may serve an effective method for finding an aggregated measure of agricultural productivity. It would be appreciated that level of agricultural development and agricultural productivity are the consequences of mainly diffusion of farm technology that has been in operation in a territory for some time. The population-potential model may, therefore, be made use of for indentifying the gravitative pulls in agricultural innovations which essentially result in the varying regional contrasts in agricultural productivity. As such this model may be profitably used for finding out productivity index. This may be done by selecting principal crops grown in the study region. The output of each crop in each base unit of the region may be determined in terms of yield per hectare, price-weighted output, per worker and SNU. This would envolve the consideration of land factor, labour factor and sufficiency factor. These values may be converted into potentials by making use of the population potential model. The different potential values for each base unit may be totalled and averaged to give a final productivity score. This method has been tested by Dube (1986) in a study of agricultural patterns in Madhya Pradesh.

References

Dondyal, S.P. (1964) : "Regional Variations in Agricultural Development and Production in Eastern and Western Regions of U.P." Ind. Jour. Agt. Eco. Vol. 19 (i), pp. 193-197.

Dube, R.S. (1984) : "Agricultural Productivity in Madhya Pradesh", Annals Nat. Assoc. Geog. India, Vol. IV (2) pp. 13-24.

Enyedi, G : "Geographical Types of Agriculture", Applied Geography in Hungary. Budapest.

Ganguly, B. N. (1938) : "Trends of Agriculture and Population in the Ganges Valley. London, p. 38.

Harring, J. (1964) : 'Concept of Productivity Measurement on a National Scale' Documentation in Food and Agriculture. Paris : O.E.C.D., p. 10.

Hirshch, H.G. (1943) : "Crop yield Index", Jour. Farm Eco. Vol. 25 (3), p. 583.

Kendall, M. G. (1939) : "The Geographic Distribution of Crop productivity in England" Jour. Royel Stat. Soc. Vol. 162, pp. 21-48.

Pandit. A.P. (1965) : "Application of Productivity Concept", Productivity Vol. 6 (2-3) p. 87.

I.S.A.E. (1964) : "Regional Variations in Agricultural Development and Productivity", Indian Jour. Agri. Eco. Vol. XI (i), pp. 168-266.

Saxon, E.A. (1965) : "Issue on Agriculture", Productivity, Vol. 6 (2-3), p. 226.

Shafi, M. (1960) : "Measurement of Agricultural Efficiency in U.P." Eco. Geog. Vol. 36 (4), pp. 296-305.

Shafi, M. (1971) : "Measurement of Food Crop Productivity in India" pp. 97-113 in M. Shafi and M. Raza (eds) Stu. in Applied Regional Geography. Aligarh : Deptt. of Geog. AMU Aligarh.

Singh, J. (1972) : "A New Technique for Measuring Agricultural Efficiency in Haryana, India", The Geographer, Vol. 19 (1) pp. 14-33.

Stamp, L.D. (1958) : "The Measurement of Land Resources" Geog. Rev. Vol. 48 (1), pp. 110-116.

Thompson, R.J. (1926) : ' The Productivity of Britain and Davish Farming", Jour. Royel Stat. Soc. Vol. 89 (2), p. 218.

Yang, W.Y. (1965) : "Methods of Farm Management Investigations for Improving Farm Productivity. Agricultural Development, Paper No. 8. Rome : FAO.

Zobel, S.P. (1950) : "On the Measurement of Productivity of Labour", Jour. Amer. Stat. Soc. Vol. 45, p. 218.

Stewart, J. Q. et al. (1958) : "Macro, geography, and social science" Geos. Rev. Vol. 48, p. 167.

CHAPTER EIGHT

AGRICULTURAL PRODUCTIVITY IN MADHYA PRADESH*

Introduction

Madhya Pradesh, is now no more a surplus producer of food crops. Though the State exports some ten million quintals of food-grains every year, it imports some four million quintals. The per hectare population supporting capacity of the crop land is only 2.40 standard nutritional units (SNU) notwithstanding the fact that a hectare of land carries some 2.88 persons. Evidently, this places the State virtually into a deficit position vis-a-vis the foodgrain output.

The population of the State is increasing rapidly. The 1981 Census enumeration reports an addition of some ten million persons against an increase of some nine million in 1971 and 6.3 million in 1961. Even if the current rate of population growth registers a moderate decline (although the possibilities are, indeed remote) in the following decades, we will have at least 20 million additional mouths in Madhya Pradesh to feed by the end of this century. Madhya Pradesh would require a good increase in agricultural productivity both for supporting its growing numbers and for regaining its surplus status in the country. A knowledge of the productivity and existing constraints on it in different parts of the State would be helpful in the formulation of a well conceived policy for improving productivity.

Boserup (1965) compares the productivity of land in the different land use systems of the world and argues that agricultural

---

* Reprinted from Annals of the National Association of Geographers, India, Vol. IV, (2), 1984, pp. 13-24.

productivity is stimulated by sustained population growth. Madhya Pradesh affords an example of a steady developing subsistence agricultural economy and exhibits a mixed pattern of the land use systems. On the one hand, there are the forest fallow systems in hill areas, e.g. Bastar, Satpura-Maikal region and the western hill borders of the State, where soil fertility is strained by the practice of shifting cultivation. At the other extreme are the areas where intensive cultivation is practised with the help of improved modern techniques. It would be interesting indeed to examine the patterm of change in the agricultural productivity of the different parts of the State and to identify the forces operating on these spatial variations.

The following paper is an attempt to interpret the regional variations of agricultural productivity in different parts of Madhya Pradesh. It seeks to identify relationships between productivity and nature of terrain, level of agricultural development, and level of social well-being. The analysis is based on district wise data and physiographic divisions has also been taken into account (Fig. 1).

Methodolgy

Four indices have been chosen : Crop Yield Index (Yang, 1965), population supporting capacity of the farm land (Stamp, 1940, Shafi, 1960), and price weighted productivity index viz. productivity per hectare (Nangia et al. 1967) and productivity per worker (Yates, 1960).

The regionalisation of the productivity indices has been based on the grouping of the districts by making use of Kendall's ranking method. The emerging productivity patterns have been correlated by making use of the Spearman's Ranking Correlation Coefficient Method for identifying inter-district variations in the level of agricultural development and those of social well-being for obtaining an understanding of the socio-economic constructions of the productivity. Kendall's method has also been used to ascertain the level of agricultural development. The developmental inputs included for this evaluation are : use of irrigation, fertilizers, tubewells irrigation pumps and tractors. The level of social well-being has been identified on the basis of the average value of Z-scores for the districts in respect of the following variables (Dube 1982).

Education

Literacy of general population, literacy of the scheduled castes and scheduled tribes; school spacing, per student expenditure on education, student-teacher ration.

Health

Population-doctor-ratio.

Economy

Population supporting capacity of the crop land (SNU); Price weighted agricultural productivity (per hectare and per worker); use of irrigation and fertilizers, per capita availability of cultivated land, landlessness, cropping intensity.

The Living Environment

Houselessness, electrification of villages, connectivity as measured by rail-road density, number of rainy days, population per commercial bank, urbanization, availability of non-agricultural occupations (gauged through the participation rate of workers in the non-agricultural activities).

Social Order

Security (Population-police ratio), political participation rate (percentage of votes polled in the legislative elections 1980), dependency ratio, level of unemployment.

The analysis is based on the 1979 agricultural data and on the 1981 census data of population. The critical limits of different classes for productivity, agricultural development and social well-being have been worked out as per cent values from the mean value in order to show the regional variations with reference to the State average.

Productivity Based on the Crop Yield Index

The Crop Yield Index (Yang, 1965) measures the ratio of the average yield of the crops of a district to the yield of those crops in the entire State.

The Index is obtained by dividing the average yield of a crop with the average yield of the same crop in the entire State and multiplying it by 100 to give it a percentage expression. The product is, then, weighted for the area under the crop. The indices of different crops are summed up and average to find the crop yield index.

Since both foodcrops and non-foods crops are involved in its computation, the Index provides a good measure of spatial variations in agricultural productivity of a region.

Twenty-one crops are taken into account in the calculation of the crop index. These are rice, wheat, jowar, maize, bajra, barley, kodo-kutki, gram, tur, moong, urd, kulthi, peas, lentil groundnut, sesamum, rape and mustard, linessed, ramtil, sugarcene and cotton. Other crops have been excluded as they are not of much significance in the agricultural economy of Madhy Pradesh.

The crop index varies from 26.7 in Khargone district in the west to 209.2 in Morena district in the north, suggesting that the range of productivity variation is as low as about one-fourth of the State average and as high as twice of it. The areas where productivity as seen from the spatial distribution of the index, is lower than the State average include twenty nine out of the forty-three districts of the State. Table 8.1 suggests that thirty districts classed under higher and lower medium categories are close to the State average. The number of districts on either side of the medium categories is distributed quite uniformly giving a normal distribution. The high and very high classes include six districts against seven districts classed under the low and very low categories (Table 8.1). Since any increase in technological innovations in the current agricultural setting would be revealed in a negatively skewed distribution of the productivity indices, the normal distribution of the crop index in Madhya Pradesh suggests that little improvements in agricultural productivity have occurred from the introduction of recent technological innovations in agriculture.

TABLE 8.1

Madhya Pradesh : Crop Yield Index, 1979

Categories	*Index*	*No. of districts*
Very low	Below 49.2	1
Low	49.2 to 73.7	6
Low medium	73.7 to 98.3	22
High medium	98.3 to 122.9	8
High	122.9 to 147.5	2
Very high	Above 147.5	4

State Average=98.3 (Based on Yang's formula)

It would be evident from map. 8.11 that the acore of the Index is on the higher side of the State average in the relatively thickly peopled tracts in the Madhya Bharat Plateau. Northern parts of the Bundelkhand Upland, eastern part of the Rewa Plateau, some districts of the western Malwa (Mandsaur, Indore, Dewas, Khandwa districts), the districts of the Upper Wainganga Basin and the Chhattisgarh basin. The thinly populated Baster, where shifting fallows still dominate the agricultural landscape, also registered a high score. The high productivity seems to be associated with (i) some areas of the fallow land use system and sparse population e.g. Bastar (ii) thickly peopled fertile valley areas, e.g. Balaghat, Bhind and Tikamgarh districts, (iii) Areas of commercialised crops, e.g. Mandsaur. The low productivity areas include parts of the Satpura Maikal-Baghelkhand region and some hilly parts of the Malwa and Madhya Bharat plateau where fallows are being rapidly reduced with the sustained growth of population.

It would seem, therefore, that productivity over a greater part of the State is moderate; for thirty districts out of the total forty-three districts are in the Lower Medium and Higher Medium categories. This indicates that nature of terrain, stage of agricultral growth and populousness of the area have a bearing on crop productivity in the different parts of the State.

Productivity Based on Population Supporting Capacity of The Cropland

The Crop Yield Index is not adequate for the assessment of the sufficiency-deficiency situation of agricultural productivity in terms of food supply; and hence the SNU approach may be more usefully employed for this purpose. Following Stamp (1959), Shafi (1960) and others for computing productivity under Indian conditions the per hectare output of foodcrops has been calculated in terms of Standard Nutritional Units (SNU). The method envolves the conversion of food production per hactare into calories, and on the basis of minimum caloric requirement of an average person, the average number of people who can depend on the food produced in the unit area are considered. Stamp termed this number as SNU. Basing our calculations on a table published by the Indian Council of Medical Research (Gopalan et al., 1980) and considering the average requirement of a person in India as 2200 calories (Shafi,

1950). the per hectare inter-district variations in SNU have been arrived at. Even though this method suffers from the weakness of excluding non-foodcrops, it provides a fairly reliable measure of productivity in the areas where foodcrops constitute major segment of agricultural produce. The principal crops taken in the calculation of SNU include rice, wheat, jowar, maize bajra, barley, kodo-kutkin, gram, tur, lentil, urd, moong, groundnut, rate and mustard, and linseed.

The per hectare yield of the crops has been taken as the basis for calculation of the SNU. Since the purpose of this study is the evaluation of the productivity of the farm land and not the foodgrain availability, only the loss on wastage between the farm and the kitchen has debited from the yield of the each crop: the loss in seed and cattle feed has been ignored. To make them more meaningful the rates of wastage through extraction, storagse, damage etc. have been estimated on the basis of field enquiry by the author in an M.P. village (Table 8.2), in preference to those adopted in the publication of FAO and by other agricultural workers (FAO, 1955, Akroyd, 1965, Ali Mohammad, 1977, Noor Mohammad, 1978). The rate of loss in extracting edible parts from the individual grains is determined by conducting laboratory measurements. A comparative statement of the present rates with those employed by earlier workers are appended in Table 8.2.

Over a greater part of the State (28 districts), the SNU is lower than the State average of 2.40. The values vary from 1.60 in the Rewa district to 7.08 in the Gwalior district. The dispersion of values on the either side of the mean is gradual, suggesting a normal distribution. Less than half the number of the districts get bunched in the proximity of the mean, and the member in the High Category is smaller (eight districts) than that in the low category (sixteen districts). This suggests that there has been little progress in increasing agricultural productivity and that a greater part of the State is still under the traditional stage of agricultural growth.

TABLE 8.2

Rates of Wastage

Crops	*FAO*	*ICMR*	*Ali Mohd.*	*Noor Mohd.*	*Results of present enquiry*
Rice	66.66	66.66	66.66	66.66	60
Wheat	90	90	90	90	97
Jowar	90	95	90	95	96
Barley	70	70	70	70	85
Bajra	90	—	90	—	96
Maize	89	89	89	89	98
Kodon	95	66.66	96	66.66	65
Gram	95	95	95	95	85
Masur	95	100	95	100	90
Urd	95	100	95	100	90
Mung	95	—	95	—	90
Arhar	95	—	95	—	85
Teora	—	—	—	—	85
Groundnut				51	
Mustard	33.33		33.33	52	
Seamum				45	
Linseed				40	

The areas of High population supporting capacity are generally those of the Madhya Bharat Plateau and some areas of the Malwa and the Narmada valley. The low productivity areas occupy extensive parts of the State in the eastern Madhya Pradesh, the Satpura and the western parts of the Malwa (Map 8.12). It would seem that regional variations of SNU in this State do not exhibit any direct relation with the degree of population concentration. The densely populated Lashkar plain in the north shows high population and high prodctivity while there is low productivity in the high population density regions of the Rewa plateau and Chhattisgarh basin. Similarly, some of the thinly populated parts of the Malwa have high productivity and low productivity in some sparsely peopled parts of the State e.g. the Baghelkhand plateau. It may also be noted that SNU based productivity does not exhibit any evident relation with the land use system of the areas. The agriculturally developed Western Malwa (Map 8.12) is classed under the low productivity categories, while agriculturally developed Madhya Bharat plateau is placed in the high productivity classes.

TABLE 8.3

Madhya Pradesh : Standard Nutritional Unit 1981

Categories	*SNU*	*No. of dists.*
Extremely Low	Below 1.68	1
Very Low	1.68 to 1.92	6
Low	1.92 to 2.16	9
Low medium	2.16 to 2.40	12
High medium	2.40 to 2.60	7
High	2.60 to 2.88	5
Very high	Above 2.88	3

Source : Derived from the district data of Madhya Pradesh, 1979.

It should follow, therefore, that the regional variations of SNU do not conform to the theory advanced by Boserup since they fail to exhibit any direct correspondence with the populousness and land use systems of the different areas of the State.

Productivity Based on per Hectare Output

The effective support of agriculture does not find its expression only in the output of farm products in terms of SNU, it can also be expressed in terms of efficiency of a piece of land in providing price rated profits to the farmers. An area may yield greater caloric output but may have smaller value of the crops produced and vice-versa. The measurement of the agricultural productivity of a region needs, therefore, to be based on the price weighted indices also. The value of farm output of all the crops raised by farmers gives a very diversified spatial patterns ranging from Rs. 762 per hectare in the Jhabua district. In the greater part of the State (thirty-three districts of the total) the values of the farm output per hectare ranges closer to the mean of Rs. 1152 (Map 8.13). The areas of Low price productivity are less (three districts), while those of high productivity occuy a relatively greater extent (seven districts). This may suggest, therefore, that the price rated productivity is moderate over greater part of the State and one may hope that rapid extension of agricultural innovations may help the expansion of the High productivity areas in the State.

TABLE 8.4

Madhya Pradesh : Price Weighted per Hectare Productivity, 1979

Categories	*Price weighted index of output (Rupees)*	*No. of dists*
Low	Below 922	3
Low medium	922 to 1152	17
High medium	1152 to 1382	16
High	1382 to 1613	4
Very high	1613 to 1843	2
Extremely High	Above 1843	1

Source : Derived from agricultural data of the districts, 1979.

Spatially, the areas south of the Son-Narmada and parts of the Malwa plateau, Bundelkhand upland and Madhya Bharat plateau have registered higher price weighted per hectare agricultural productivity, suggesting its fairly close correspondance with the nature of terrain. The relatively undulating parts of the State in the areas north of the Satpura-Maikal ranges have registered lower productivity. It seems that the nature of terrain in combination with populousness and stage of agricultural growth determine the productivity status of the districts, since some correlation of productivity with population concentration and agricultural development is descernible from Map 8.13.

Thus, productivity over a greater part of the State is moderate in respect of the price weighted aerial productivity also.

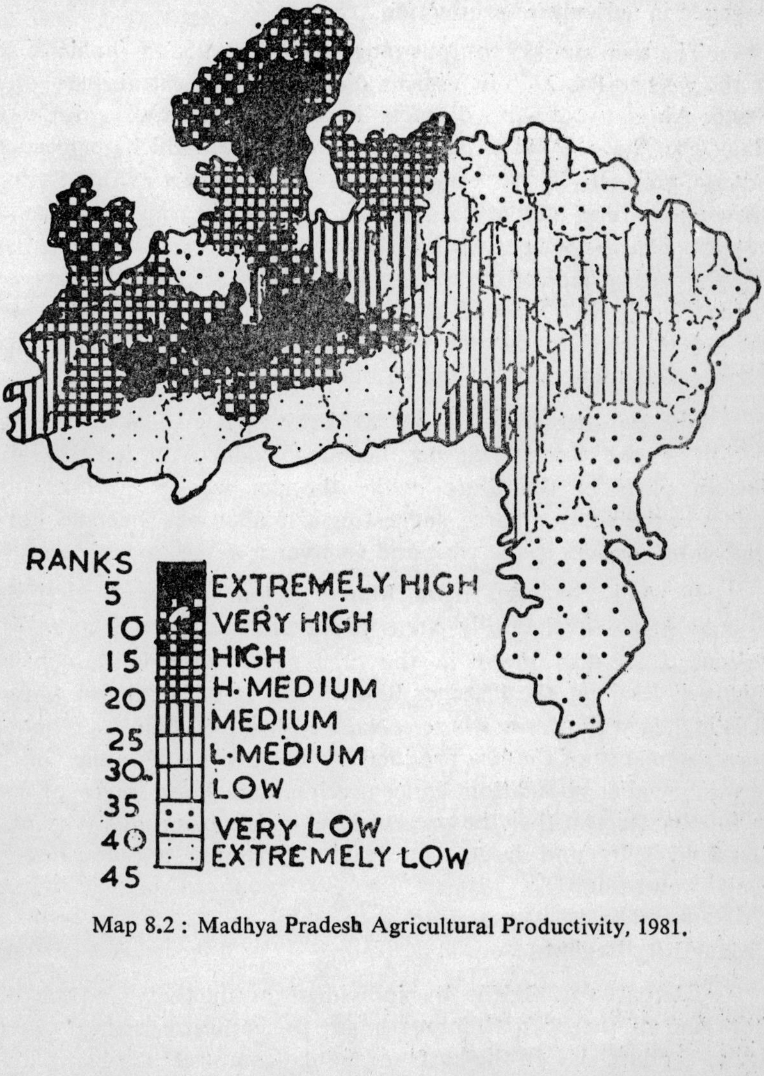

Map 8.2 : Madhya Pradesh Agricultural Productivity, 1981.

Productivity Based on Output Per Worker

The price wighted productivity of crops per agricultural worker is useful as it is a good indication of the income obtained by farmers engaged in agriculture production.

The per worker output ranges from Rs. 955 in Jhabua distrist in the west to Rs. 2755 in Vidisha district in the central part of the State. About twenty-six districts have recorded lower productivity than the State average of Rs. 1758 per year, which suggests that labour productivity is very low over a large part cf the State and that agricultural activity is still at an initial stage of agricultural growth. The output per worker shows a normal type of distribution on the either side of the mean. The High productivity areas (comprising nine districts) extend in the Malwa and Lashkar plains, while the low productivity districts (twenty-six) are confined to the Rewa Plateau and some areas of the Satpuras.

The per hectare and per worker output give an inverse distribution in the State. The per hectare productivity is higher in the eastern parts of the State, while the per worker productivity is higher in the western parts, suggesting a smaller per hectare concentration of workers in the west and vice-versa in the east (Map 8.14).

It would be seen from Map 8.14 that labour productivity is usually lower in the hilly areas of the State, e.g. the Satpura-Maikl region. It is also lower in the areas where population density is relatively less, e.g. the Plateaus Bastar areas of Baghelkhand, Shivpuri. It is higher in the areas where social well-being is higher. The close correspondance of the low productivity belts with the areas of hilly terrain, smaller population concentration and lower level of social well-being suggest that the present level of labour productivity of the farmland is, by and large, the result of terrain populousness and social well-being.

Productivity Regions

The patterns shown by individual productivity indices differ from one another. An attempt is made to synthesize spatial variability in productivity on the basis of some composit scores following Kendall's method (1939) The ranking of the districts for each of the indice has been taken in descending order. The ranks so obtained are, then, totalled and averaged to yield composit score for each district. This is used agein for final ranking of the districts. The first ranking district (i.e. the one with the highest score) has the lowest productivity and the last ranking (i.e. the one with lowest score) has the highest. The ranked districts are grouped into six classes : Very low Medium, High Medium, High and Very High. These are, then, mapped to show the regional variations in productivity and to delineate the productivity regions of the State (Map 8.2).

The spatial variability in productivity seems to have an apparent association with phyical character of the terrain. The rolling upland plains in the Madhya Bharat, Bundelkhand and Chhattisgarh show High productivity as against Low productivity in the intensively dissected hilly Satpuras and Bundelkhand. It nevertheless seems to have some relation with non-physical factors ; the overall degree of correlation with the agricultural development (P = 0.46) and the social well-being (P=0.44) is insignificant. Any search for a rational *inter se* relationship would necessitate the examination of productivity by individual productivity regions of the State. The following six regions are identified in Fig. 7—1. The Malwa Plateau, 2. The Madhya Bharat Plateau ; 3. The Bundelkhand Upland, 4. The Narmada valley; 5. The Vindhyachal Baghelkhand Region, 6. The Chhatisgarh region.

The Malwa Plateau

The Malwa registers a mixed pattern of agricultural productivity. It ranges from very Low to very High, and there are no compact extensive belts of Low or High productivity areas in this region. Agricultural productivity in this part of the State seems to hold closer links with the nature of the terrain since value of Spearman's Correlation Coefficient (P) with the non-physical variables. viz. the agricultural development (P=0.08) and the social well-being (P=0.48) is insignificant.

The pressure of population on the land is lesser and the size of a agricultural work force is relatively smaller in this part of the State. The physiological density of population usually does not rise more than the High Medium and per capita cultivated land is rarely less than half the hectare. The lesser population pressure and paucity of the agricultural labour have perhaps combined to slow down agricultural growth in this region, and productivity seems to have a closer correspondance with the physical character of the land than with agricultural development and social well-being. In fact the stage of agricultural growth, where social well-being and agricultural development combine to stimulate agricultural productivity for satisfying the increasing demand of farm products, has not yet come in the Malwa.

The Madhya Bharat Plateau

This is one of the best productive parts of the State, productivity here varies from Low in the Khichiwara Plateau in the south to very High in the Lashkar plain in the north (Map 8.2). The per hectare productivity in respect of Crop Index and population supporting capacity (Map 8.11 and 8.12) is high as against price weighted areal and labour productivity (Map 8.13 and 8.14).

Though, productivity evinces a close relationship with the terrain. it also shows a strong correlation with agricultural development (P=0.85) and social well-being (P=0.79). This is a relatively more densely populated area of the State, and per capita cultivated land is scarcely more than half hectare in the more populous parts of the region. The smaller availability of agricultural land and the growidg pressure of population have caused need for better agricultural growth. The higher level of social well-being of people in this region implies possibly better development inputs. and hence agricultural productivity yields a strong correlation with both the non-physical variables selected. Incidentally, it may be pointed out here that there is still a greater emphasis on increasing yields rather than on raising the higher value crops. Any future increase in the population pressure is likely to shift this emphasis towards higher value crops.

The Bundelkhand Uplands

These upland plains of medium population density, register Medium to High productivity. Productivity is moderate in respeet of all the indices except Crop Index (Map 8.11 and 8.13). Since the plateau soils are often poor and have smaller moisture retentive capacity, and since per capita cultivated land is smaller (less than half hectare), the land demands more intensive care for cultivation and necessitates a more frequent use of developmental inputs for increaing farm output. Productivity gives a strong correlation with the agricultural development (P=0.80) though there is feeble correlation with the social well-being (P=0.40).

The Narmada Valley

The Narmada valley has a better accessibility on account of the passage of one of the busiest railway routes through this region. The physiclogical population density over a greater part is Medium and per capita cultivated land ranges around half a hectare.

The pattern of productivity in the region is a mixed one (Map 8.2). Though the population supporting capacity is High (Map 8.12) the aerial and labour productivity is of lesser order (Map 8.11. and 8.13). The emphasis of cultivation seems to be more on nutritive crops rather than on higher price crops. The degree of correlation with agricultural development (P=O.28) and social well-being (P=O.O7) is, indeed, insignificant. Since the land is suitable for cultivation and the dependence of people on land is less, to some extent, because of non-agricultural activities, the need for more intensive use of soils has not yet emerged. Consequently forces of social well-being and agricultural develoment have yet to influence agricultural productivity in this region.

The Vindhyachal-Baghelkhand Region

In this region physiological density is Medium though there are some exceptions. Rewa district, for instance, has high density of population and per capita cultivated land is less than half a hectare.

The region has Low to Very Low productivity. Sinece the availability of agricultural land is limited and the soi1s are less fertile, more intensive agriculture is necessary for increasing capacity of the garicultural land, Poor connectivity inhibits application and diffusion of efficient agricultural technology and any better use of the soils in such a situation would depend, therefore, primarily on the social well-being of the farmers. This is why agricultural productivity shows a closer co-respondance with the level of social well-being (P=O.83) in this region notwithstanding insignificant correlation with agricultural development.

The Chhattisgarh Region

This rice bowl of Madh a Pradesh, shows a mixed pattern of agricultural productivity varying from Low Medium in the north to Very High in the south (Map 8.2). Though aerial productivity is

High (Map 8.12 and 8.13), population supporting capacity and labour productivity are lower (Map 8.12 and 8.14). With the solitary exception of the Bastar district in the south, this is one of the high density regions of the State and per capita availability of cultivated land is less than O.40 hectare.

The High productivity in the Dandkranya (Bastar district) ignores all the dictates of productivity—the terrain, populousness agricultural development and social well-being. Thisis an intensively dissected hilly and forested part of the State. It is at the lowest level in regard to the population density, agricultural development and social well-being (Map 8.15 and 8.16). The agricultural system is at the initial phases of agricultural growte and is dominated by forest fallows. The High productivity seems to have a closer link with the high fertility of forest soils stimulated by shifting rotation of forest fallows.

The rest of the Chhattisgarh shows a moderate level of agricultural development (Map 8.15), but High level of social of well-being (Map 8.16). It compares the Madhya Bharat region of the State in respect of conditions of agricultural growth, since it gives a strong positive correlation with both of the variables, viz. the agricultural development (P=O.98) and social well-being (P=O.66). There is a greater emphasis here on high price crops (Map 8.13), which indicates that the population pressure has resulted in the resulted in the shift of cultivation towards higher price crops.

The Emergent Trends

The emergent relationships of the productivity in Madhya Pradesh with population pressure, stage of agricultural growth, nature of terrain, level of agricultural, development and social well-being, presented as a model are as follows :

1. The high productivity areas in the State are generally associated with the valleys and upland plains of the Madeya Bharat plateau, the Malwa Plateau, and the southern parts of the Chhattisgarh region. The low productivity areas generally extend into the hilly and dissected areas of the Vindhyachal Bagelkhand region and those of the western and northern fringes of the Malwa.

2. There is no evident *inter se* spatial correspondance between agricultural developmnt and social well-being, they seem to hold variable correlation with agricultural produtivity.

3. A strong correlation of productivity with agricultural development and social well-being seems to be associated with the areas where pressure of population is high and where the growing need for greater farm output has necessitated adoplion of agricultural development inputs. These, in their tnre, are largely determined by the level of social well-being of the farmers. Such is the situation in the belts of the Madhya Bhart Plateau and the Chhattisgarh basin.

4. The areas where significant correlation emerges with social well-being alone are those parts of the State where nature of the terrain is not much suitable for agricultural growth. Rugged relief and poor connectivity have largely acted as barriers against development technology and smaller population pressure has failed to generate forces which invigorate the process of development.

5. The areas where there is a strong correlation between productivity and agricultural development. ignoring the influence of social well-being for the adoption of the agricultural development inputs, are those where poor soils demand intensive cultivation to support the people dependent upon them. The Bundelkhand upland examplifies such an area where irrespective of social well-being of the farmers there is diffusion of new agricultural techniques owing to its proximity with certain areas of the more developed Gangetic plains.

6. The areas of the Malwa and the Narmada valley show no strong relationship between productivity, agricultural development and social well-being. There are the areas where land is favourable for agriculture and population pressure is small, and a strong need for the adoption of agricultural innovations and techniques motivated by social well-being is yet to be generated. Productivity, shows therefore, a greater correlation with the nature of land and none with the other two variables.

7. It follows, thus, that the kind of relationship of paoductivity between the terrain, agricultural development and social well-being would depend upon the populousness of a region and the stage of agricultural growth. In the initial phases of agricultural growth

where forest fallows dominate land-use and population sparse, adequate agricultural production would be possible irrespective of any change in agricultural techniques and social well-being of the people. This is the case in the Bastar region of the Madhya Pradesh, where productivity shows close correspondance with the nature of the terrian. In the subsequent phase, when density increases and 'fallow period' is shortened to cause a state of diminishing returns from the land. productivity would vary with level of social well-being for, in the absence of any availability of the development inputs, the better off people would tend to get better yields from the farm than those of the relatively poorer and distressed communities. This is exemplified in the Satpura-Maikal Baghelkhand region of the state where lower level of social well-being has exerted its influence to keep productivity level lower. In the later stages, however, sustained population growth and resultant increase in poputation density would substantially reduce fallows and soil fertility, and in the absence of any agricultural innovations productivity would decline substantially In the Rewa region in northern part of the State, for instance, agricultural productivity is lower, though the terrain is favourable of social well-being is higher. The agricultural development seems to have failed there to keep pace with the population growth. The growing demand for agricultural production would, therefore, subsequently introduce changes in agcultural techniques for raising agricultural productivity. as is the case in the Chhattisgarh and the Madhya Bharat Plateau. The adoption of techniques and innovations would, however, essentially depend on the capacity of the farmers to respond to these innovations and techniques. Consequently, agricultural productivity in this phase of agricultural growth in a subsistence economy would vary in direct relation with the level of agricultural development and that of social well-being.

The regional associations and the emergent trends in the agricultural productivity Madhya Pradesh, as stated above, are similar to those advanced by Boserup (1965). Boserup hasm, however, ignored the consideration of the role of social well-being in the growth of the agricultural productivity, notwithstanding the fact that it plays a vital role in the decision making process of farmers for adoption of developmental inputs. The model of agricultural productivity as presented in this paper differs from that of Boserup's Boserup has treated level of productivity as an adaptation to population growth, while the model of agricultural productivity adopted here is presented as an out come of nature of terrain, stage of agricultural growth, and level of social well-being which exerts its influence throngh the adaptability of agricultural innovations.

CONCLUSION

The regional aspects of agriaultural productivity studied as inter-district variations of Crop Yield Index, SNU, and price weighted productivity sharp contrasts in Madhya Pradesh.

REFRENCES

Akroyd, W.R. and Gopalan, C. (1965), The Nutritive value of Indian Food and Planning of Satisfactory Diet. Indian Councial of Medical Research, New Delhi.

Boserup, E. (1965). The conditions of Agricultural Growth. Allen & Unwin, London.

Chakrawarti, A.K. (1970). Foodgrain Sufficiency Patterns in India. Geog. Rev. Vol. 60 (2), p. 217.

Directorate of Agriculture, Madhya Pradesh (1979) : Agricultural Statistics of Madhya Pradesh, Bhopal.

Dube, K.C. (1981). Census of India, 1981. Series II ? Madhya Pradesh Provisional Tables. supplement : Paper 1. Directorate of Census Operation, Bhopal.

Dube, R.S. (1979). Population of Rewa Plateau. Sahitya Ratnalaya, Kanpur.

Dube, R.S. (1982). "Social well-beiag in Madhya Pradesh". Trans. Inst. Ind. Geog. Vol. 4(2), pp. 169-174.

FAO (1955) : State of Food and Agriculture, Rome.

Gopalan, C. et al. (1980) : Nutritive Value of Indian Foods and Planning for Satisfactory Diet'. Indian Council of Medical Research, New Delhi.

Kendall, M.G. (1939). "The Geographtc Distribution of Crop Productivity". Jour. Royal Stat. Soc., Vol. 162, pp. 21. 48.

Mitra, A. et al. (1980). Population, Food and Land Inequality in India. 1971. Allied Publishers, Bombay.

Mohammad, A. (1977). Food and Nutrttion in India, K.B. Publications New Delhi.

Mohammad, N. (1978), Agriculture and Land Use in India, Inter India Publication, Delhi,

Nangia, S. et al (1967). "Variations in Field Productivity : A case Study of Khandawala, Haryana", The Geographer, Vol. 14, pp. 12-14.

Shafi, M. (1960). Measurement of Agricultural Efflcieucy in Uttar Pradesh. Econ. Geog., Vol. 36(4), pp. 296-305.

Singh, J. (1972) : A New Technique for Measuring Agricultural Efflciency in Haryana, India". The Geographer, Vol. 19 (1), pp. 14-33.

Stamp, L.D. (1958) : Measurement of Land Resourles" Geog. Rev. Vol. 48 (1) : pp. 110-116,

Yang, W.Y. (1965) : Methods of Farm Management Investigatians for Improving Farm Productivity". FAO Agricultural Development, Paper No. 80, Rome.

Yates, P.L. (1960) : Food, Land Manpower in Western Europe. pp. 147-163.

CHAPTER NINE

EVALUATION OF FOOD SUFFICENCY

Introduction

The evalution of the dimensions of sufficiency of food production in a territory is one of the fundamental tasks of the agricultural geographer. The knowledge of level of self sufficiency in respect of food products is important for the formulation of a well-concieved food strategy and for the implementation of a rational distributive network of the circulation of foodstuffs in an area. The agencies concerned with the responsibility of the measurement of the food supply system must, therefore, have full knowledge of the surplus-deficit situations of the food production before taking steps for enforcing an effective circulatory system of food stuffs.

The Connotations

The term food sufficiency refers to the level of human stisfaction in respect of the production of food stuffs in the areas occupied by a group of human individuals. The term is used as a synonym of the 'food availability' (Dube and Misra, 1984) or 'Net Food' (Ali Mohammad, 1977 and FAO). It differs from the terms food productivity and food consumption. The productivity refers, as explained in Champter 7, to the power of a productive system to provide products for human use and therefore relates the production with the elements envolved in its production, e.g. farm, labour, machinery etc. The connotation of the food productivity therefore, measures the amount of food stuffs produced per unit farmland or per worker engaged in its production. This is expressed either as weight of the products or the nutritive/money value derived therefrom, The lovel of food consumption refers to the power of the human individuals of a community to consume the available food products. It affords the estimation of the amount of foodstuffs which actually find access to human diet. They may be produced locally or may be imported from elsewhere. The determination of the level of food intake in a territory may, therefore, depend on the data obtained from food surveys or from food balance sheets prepared after giving allowance for export and imported food stuffs.

The determination of food sufficiency, coversely relates the food stuffs produced in a territory during a given period of time with the individuals of the human population, living in that territory. It expresses the amount of the food products which becomes available for human consumption in the edible form.

Determination of Sufficiency

There are two steps in the determination of the level of food sufficiency in areas :

1. Estimation of food products from different sources,
2. Estimation of the amount of food, which may be available for human consumption, locally or elsewhere.

Estimation of Food Products

Determination of 'food availability' in a greater number of regional studies affords the estimates of only partial availability, since they consider only the foodgrains and do not take into account the food available from the sources other than the food crops. This is so partly because food of a greater segment of the world population is based primarily on the cereals and pulses, and partly because of the pauciety of data regarding the other sources of food products. Any evaluation of the food sufficiency of a region must, however, depend on the determination amount of food stuffs from all thesources that supply food for human subsistence.

The principal diffieulty in the determination of food supply from all the sources of food is the paucity of data regarding foodstuffs obtained from the sources other than the food crops. Data regarding the food supplies from such sources as animal meat, eggs, milk, fishes etc. are rarely available in statistical records of different agencies and production of these stuffs would be required to be estimated from indirect sources of data. This author is of the view that available statistics on the populations of animals which provide food products may be taken as the basis for estimating their food yielding capacities,

The first step in this task would be the identification of the animals of the territory under question which provide such food products as meat, milk and eggs. Since all meat yielding animals can not be slaughtered for meat and since all the milk or egg yielding animals are not always in the lactating or egg producing conditions, the frequency of such animals would have to be fixed before making any estimates for those products. The next step would, then, be the determination of average yield of meat, milk or egg from the concerned animal, which could be ascertain by conducting small local surveys. The total availability of these products would then be computed taking into account the meat, milk/egg yielding sections of the concerned animal growps. Misra (1982) has attempted to provide an estimate of yearly supplies of meat and eggs in Madhya Pradesh on the assumption that about 75 percent of goat, sheep and cocks are slaughtered for meat and also that average goat, sheep and cocks yield 15 kilograms, 10 kilograms one kilogram of meat respectively. He also estimates milk production on the assumption that an average cow and an average buffalow yield about 1.5 and 3.5 litres of milk respectivaly and on the assumption that their average lactation period range in the neighbourhood of 240 days and 300 days respectively. He conjectures that of the total populatiod 74 percent of cows and 79 percent of buffalows may be taken as milk yielding. His assumptions may be dubious but it may afford a yardstick for the evaluation of food supplies from those sources, data for which are not registered by records keeping agencies of a region.

Determination of Foodgrain Availability

Foodcrop production of an area is invariably subjected to the losses arising from animal feed, seed, manufacture, extraction, and wastages (FAO, 1957), Any estimation of the amount available for human consumption would therefore involve debiting the quantities that fail to be used for human consumption. The desirability of invelving consideration of the export-import component of grain mobiiity (FAO, 1957; Ali Mohmmad, 1977), however, seems dubious. Firstly, the data on movent of grains are difficult to assemble at the micro level in India. Secondly foodgrain mobility does not involve any permanent loss of the grains except of small quantities which may be lost during their movement from one area to another. In fact, export-import is a kind of human arrangement for affecting the distribution and regulation of grain from surplus to deficit areas to save destruction of the grains from their non-utilisation. Thirdly, if the basic purpose of measurement of availability is knowledge of the surplus-deficit position of the areas, the export-import factor would hardly have any relevance in availability estimates. The inclusion of the export-import factor would, in fact, indicate the amount of grains that might be actually consumed by the people and would therefore provide a measure of the consumption and not that of the availablity. Fourthly, grain mability hardly gives any reliable indication of the surplus-deficit situation of the areas because of the fact that it may be operating under compelling circumstances resulting in a kind of districts trade (Knight, 1954; Chakravarti, 1970). The measurment of availability must therefore be essentially based on the idea of determining the quantity of grains available in consumable from irrespective of the fact that this may be consumed locally or elsewhere Dube and Misra, 1984).

Estimation of Grain Losses

There are two mithods for the ditermination of quantities of grains which fail to find accoss to human diet :

1. The one in which some standard rate of loss is adopted on some hypothetical assumption. Akroyed (1941), Stamp (1958), and Shafi (1960) and Gopalan (1980) have suggested that an allowance of 10 per cent must be made in favour of losses arising from seed, cattle feed extraction, wastages etc. Knight (1954) has however, employed the rate of 12.5 per cent for determination of the amount of food that may be lost before finding access to the Kitchen, Chakravarty has preferred the use of a rate of 16.5 per cent from seven principal crops of India, *viz* wheat, rice, bajra, barley, maize, jowar and ragi. Dube and Misra (1982) wonder that these averages may be under-estimations, since a survey conducted by them has informed the occurence of losses at the higher level (Table 9.1). Other surveys have also shown the post-harvest losses in the range of 20-40 per cent (F A O, 1974).

2. The second method is based on the construction of food-balance sheets where separate rate losses are employed for all the individual grains (Ali Mohammad, 1977, Noor Mohammad, 1978). The idea behind this approach is the fact that rates of losses for different grains differ so much that any deduction from the total foodgrains at any average rate may produce misleading estimate of the availability of food grains. This may be so specially in such

areas where there are conspicuous differences in the composition of different crops in the farms. Dube and Misra (1984) have shown that loss of grains between farm and the Kitchen differ from 6.88 per cent in case of bajra to 67.92 per cent in case of linsed in Madhya Pradesh, though the average percentage of loss for all the grains is estimated to be only 20.39 (Table 9.3). The estimation of the amount of food available for human consumption in such areas where there are marke variation in the crop types must therefore rest on the grains lost from all the individual crops at their individual rates.

Dube and Misra (1984) have tried to determine the rates of losses from different crops by conducting a field survey the result of of which are summarised in Table 9.1 and 9.2. They have fixed rates of losses acruecing both seeds and in the process of extraction of edible parts from the grains. The results of their investigation are summarised in the following pages.

Seed Rates :

The view of different workers are quite varied in respect of the allowance to be made for loss of grains for seeds. The Famine Inquiry Commssion of India in 1800 and 1900 adopted a rate of 6 percent (Singh, 1947) while Kinght (1954) and Chakravarty (1970) have adopted the rates of 2.5 percent and 6.8 percent respectively for all the crops raised their study areas. The findings of the Dube-Misra survey (1984), however, reveal that any average rate adopted for all the crops combined together would offord misleading results vis-a-vis the estimation of foodgrains availability in a territory. The percentage of average loss of grains for seeds has been shown to vary from 5.00 Kilogram per hectare (Table 9.6) in the jowar-wheat growing Shajapur district of Madhya Pradesh to 7.16 kilogram per hectare (Table 9.5) in the rice-teora growing Raipur district of the same state. In the wheat groing Sagar distsict of the state the average seed rate is shown to be 12.42 kilograms per hectare suggesting that rate of loss of grains in seeds would depend on the kind of crops that are grown in a region and that any universal rate for all the earth region is bound to produce dubious estimate of the availability. The Dube-Misra survey informs that seed rates differ from 5 kilograms per hectare for bajra to 110 kilograms for lentil (Table 9.1), The rates are shown to average 37 kilograms for cereals, 48 kilograms for pulses and 34 kilograms for oilseeds. In view of such a marked variation in the seed rates for dtfferent crops the adoption of any over all

TABLE 9.1

SEED KAVESS (Kilograms per hectare)

Cross	Jnau	Ali Mohd.	Noor Mohd.	Dube & Misro
	1	2	3	4
1. Rice	40-50	40	120	45
2. Wheat	100	110	160	100
3. Jowar	10-12	15	—	12
4. Bajra	5	—	—	5
5. Barley	75	115	120	75
6. Maize	15-20	20	10	15
7. Kodon-kutki	10-12	10	20	10
8. Gram	75	115	20	75
9. Masur	30	—	30	110
13. Urd	15-20	10	—	15
12. Mung	15-20	—	—	15
12. Arhar	40-50	10	—	15
13. Teora	—	—	—	75
14. Groundnut	100	—	—	100
15. Mustard	6	—	—	6
16. Sesamum	25-30	—	—	27
17. Linseed	3	—	—	3

Source :

1. Jawaharlal Nehru Agricultural University (4980) Krishi Viswa, Vol. 2, 3 and 4 (Jabalpur : Jawaharlal Nehru Agricultural University).
2. Calculatsd from the seed and area data of crops given in "Food and Nutrition in India", byAli Mahammad (1977), pp. 57-III (New Delhi K. B. Publications).
3. Calculated from the seed and area data of c.ops asgivn in Agricvltural Land Use in India by Noor Mohammad (1966) pp 92-172 (Delhi Inter India Publications).
4. R. S. Dube and R. P. Misra, "Evalyation of Foodgrasn Availabitity in India : An Altirnati Approach,' Asian G-agrapher, Vol. 3 (1), 1984, P.'53.

TABLE 9.2

Extraction Rates (Percentage)

Crops	*Fao*[1]	*Icmr*[2]	*Ali mohd*[3]	*Noor Mohd*[4]	*Dude and Misra*[5]		
					Extraction	*Wastage damages*	*Total*
Rice	66.56	66.66	66.66	66.66	95 (half polished	5	90
Wheat	90	90	90	90	97	7	90
Jowar	90	90	90	95	96	3	93
Bajra	90	—	90	—	96	2	94
Barley	70	70	70	70	85	2	93
Maize	89	89	89	891	98	5	93
Kodon	95	66.66	95	66.66	65	2	63
Gram	95	95	95	95	85	5	80
Masur	95	100	95	100	90	5	85
Urd	95	100	95	100	90	3	87
Uung	95	—	95	—	90	3	87
Arhar	95	—	95	—	85	3	82
Teora	—	—	—	—	85	3	82

Groundnut	—)	—)			51	48
	)	)			)	
Mustard	—)	—)			52)	49
		33.33				
Sesamum	—)	—)	33.33	45)		42
	)	)		)		
Linseed	—)	—)		40)		37

Source :

1. Fao (1955) : State of Food and Agriculture, Rome.
2. Aykroyd, W.R. Revised by Gopalan, S. (1965) The Nutritivs Value of Indian Focd and Planning of a Satisfactory Diet, ICMR, New Delhi.
3. Ali Mohd. (1977) : Food and Nutrition in India, pp, 57-III, (Ncw Delhi - K.B. Publications).
4. Noor Mohammad (1978) : Agricultural Land use in India, pp. 92-172 (Delhi : Inter-I ndia Publications).
5. R.S. Dube and R.P. Misra, *op. cit.*, 1984 p. 54.

overage rate for all the crops combined together woulp not be justified. The allowance for seeds would have to be made therefore for all the crops at their individual rates as recorded in Table 9.1.

It may be medtioned that the seed rates estimated by Dube and Misra (1984) do not compre whith those adopted by other earlier works (FAO, 1957, Ali Mohammad, 1977, Noor Mohammad, 1978). They are, however, nearer to the estimates given by J.L.N. Agricultural University, Jabalpur (1980). The reasons for marked differences in the seed rates preferred by different agencies and research workers are not easily understood, It is possible that these difference may be due to regional characteristics of the areas the products of which these estimates have been derived from.

Extraction Rates :

The rates loss of grain in the process of extraction of edible parts therefrom have also been variously adopted as shwon in Table 9.2 A marked variation in the extraction rates for different crops has been confirmed from the Dube-Misra survey (1984). It has been shown that average loss of graihs from all the crops in Madhya Pradesh averages at about 12 per cent notwithstanding the fact that average loss is 10 per cent in the rice produce Raipur district and 12 per cent in the jowarwheat grwoing Shajapur district of State. The extraction rates for the individual crops as summariesed in Table 9.2 varies from 7 per cent for Maize to 67 per cent for linseed. The rates preferred by other agencies, however, seem to be very much different from these findings. The rates given by FAO are over-estimed for cereals and under-estimated for pulses.

Losses from Other Sources

Apart from the loss of grains involved in seed and extraction some amount is also supposed to be lost in cattle feeding, manufacturing and damges occurring all along the line through poor management harvesting, in seeds, rodent. fungus, storage, poor quality of sacks and containers, and inexperienced handling in transport. The data collected in the present survey, however, do not suggest the losses through these sources as those visualised by previous workers. The edible part of the grains, for examle, is rarely allowed for cattle feeding in the territory surveyed, viz. Sagar district, specially when there is wideprad foodgrain deficiency in the country. The inedible

TABLE 9.3

MADHYA PRADESH FOOD BALANCE SHEET (Average year 1975-76 to 1979-80)
(Million Kilograms)

Crops	*Produce-tion*	*Seed*	*Gross Food*	*Extrac-tion*	*Net Food availability*	*Net loss*	*Percent loss*
Rice	3484.9	211.9	3273.0	327.3	2945.7	539.2	15.47
Wheat	2863.0	308.4	2524.6	252.5	2772.1	590.9	20.64
Jowar	1305.9	23.6	1282.3	89.8	1192.6	113.4	8.68
Bajra	97.5	9.0	96.6	5.8	90.8	6.7	6.88
Barley	168.3	14.2	154.1	26.2	127.9	40.4	23.97
Maize	649.5	10.4	639.1	44.7	594.3	55.2	8.49
Kodon	241.8	12.9	228.9	84.7	144.2	97.6	40.37
Total Cereals	8011.0	612.4	8198.6	820.9	7367.6	1443.3	22.79
Gram	1036.0	144.4	891.5	178.3	713.2	322.7	31.15
Masur	109.8	33.9	75.9	11.4	64.5	45.3	41.26
Urd	158.6	11.7	146.9	19.1	127.8	30.8	19.45
Mung	59.2	3.4	55.8	7.3	48.5	10.7	18.05
Arhar	323.6	7.5	316.1	56.9	259.2	64.4	19.90
Teora	218.6	52.5	136.0	29.9	166.2	82.4	37.71

Total Pulses	1905.7	253.6	1652.2	302.8	1349.3	556.4	27.92
Groundnut	202.0	43.6	158.5	82.4	76.1	126.0	62.36
Rape & Mustard	70.9	0.6	70.3	35.9	34.5	36.5	51.42
Sesamum	32.6	1.5	31.1	18.0	13.1	19.5	59.92
Linseed	137.7	18.3	119.4	75.2	44.2	93.5	67.92
Total 1 Oilseeds	443.2	64.0	379.2	211.5	167.7	275.5	60.41
Grand Total	11156.9	930.0	10230.0	1345.2	8884.7	2275.2	20.39
Per cent Total Output	100	8.34	91.66	12.06	79.63	24.37	20.37

Source : R.S. Dube and R.P. Misra, *op. cit.*, 1984, p. 56.

TABLE 9.4

SAGAR DUSTRICT : FOOD BALANCE SHEET (Thous and ktlograms)

Crops	*Produce-tion*	*Seed*	*Gross food*	*Extrac-tion*	*Net food*	*Net Loss*
Wheat	193100	24992.0	169108.0	16910.8	152197.2	41902.8
Rice	8500	692.1	7807.9	780.9	7027.1	1472.9
Jowar	20300	280.1	20019.9	1401.4	18618.5	1681.5
Bajra	50	0.5	49.5	3.0	46.5	3.5
Barley	1000	73.5	926.5	157.5	769.0	231.0
Maize	4300	62.7	4237.3	296.6	4870.7	359.3
Kodon	1400	63.4	1336.6	494.5	842.1	557.9
Gram	28100	4147.5	23952.5	4790.5	19162.0	8938.0
Masur	16500	4540.8	11959.2	1793.9	10165.3	6334.7
Urd	600	48.3	551.7	71.7	480.0	120.0
Mung	500	35.1	464.9	60.4	404.5	95.5
Arhar	2800	67.8	2732.2	491.8	2240.4	559.6
Teora	1040	192.0	848.0	152.6	696.4	344.6

Groundnut	4500	1132.0	3368.0	1751.5	1616.6	3383.4
Mustard	500	3.1	496.9	253.4	243.5	256.5
Sesamum	200	12.0	188.0	109.0	79.0	121.0
Linseed	6200	59.4	6140.6	3868.6	2262.0	3928.0
Total	291040	36147.0	254187.8	33388.0	221728.7	69790.2
Percent of Total output	100	12.42%	87.34%	11.5%	76.18%	23.98%

Source : R.S. Dube and R.P. Misra, *op. cit.*, 1984, p. 57.

part separated from the grains of the edible part destroyed during the course of extraction process is, in fact, allowed for cattle consumption, and there is no separate alloction for their feeding from the consumable (human) graing. Losses from the other sources also do not constitute a very large proportion of the production. In fact, the use of pesticides and efficient transportion now effectively prevent losses arising from damages of grains in storage and movement.

Since accurate data on damages are difficult to gather, we have to depend on guess work and common knowledge. Previous workers haue not listed the rates of damages separately. The present authors have, however, estimated the magnitude of damages in the range of 2 to 7 per cent deqending upon the parishability and accessibility of insects, etc. to them. There are some crops, like kodon, which can be stored for years without any damage whatsoever from the insects etc.

The Relevance of Average And Individual Rates

The usefulness of the adoption of individual rates of losses for different crops has already been discussed in the preceding sections of this chapter. The case of Madhya Pradesh is presented to show how the average rates may lead to overestimation of available food supplies in a region. An observation of Table 3 would reveal that about 20.39 per cent of the total production would be lost if allowance for the losses is affected at the rates fixed by Dun, and Misra (1984). These averages have been worked out by summing the seed rates fixed for different crops and dividing the sum by the numbe of crops involved. Dune and Misra (1984) showthat deteronation of total loss of grains on seed and then find out its per cent value with reference to the total output of the grains would be a better approach.

Dube and Misra have shown that the losses of grain from all sources vary from 6.68 per cent in case of bajra to 67.92 per cent in that of linseed. The rate of loss from cereals averages at 16.38 per cent, from pulses at 29.24 per cent, and from oil seeds at 62.16 per cent. The average rate of loss from cereals in Madhya Pradesh confirms the rate adopted by Chakravarti (1970) since he considered the availability of cereals only and did not include pulses and oilseeds in his presentation on the surplus/deficit position in respect of

TABLE 9.5

RAIPUR DISTRICT : FOOD BALANCE SHEET

(Thousand kilogroms)

Crops	*Produce-tion*	*Seed*	*Gross food*	*Extrac-tion*	*Net food*	*Net loss*
Wheat	14600	1592.0	13008.0	1300.8	11707.2	2892.8
Rice	674400	35280.9	639119.1	63911.9	475207.2	99192.8
Jowar	500	69.6	430.4	30.1	400.3	99.7
Bajra	—	—	—	—	—	—
Barley	60	6.0	54.0	9.2	44.8	15.2
Maize	100	14.4	885.6	62.2	823.6	76.4
Kodon	7700	595.0	7105.0	2629.9	4476.2	3223.9
Gram	1500	259.5	1240.5	248.1	992.4	507.6
Masur	1020	487.2	632.8	94.9	537.9	482.1
Urd	4520	383.7	4136.3	537.7	3597.6	921.4
Mung	1100	96.0	1004.0	130.5	873.5	226.5
Arhar	2300	57.3	2242.7	403.7	1839.0	461.0
Teora	54440	14103.0	40337.0	7260.7	33067.3	21363.7

Groundnut	3450	716.0	2734.0	1421.7	1312.3	2137.7
Mustard	100	9.8	99.2	50.6	48.7	51.4
Sesamum	700	39.0	661.0	383.4	277.6	422.4
Linseed	5100	1703.1	3376.8	2140.0	1256.8	3843.2
Total	772390	55303.0	717086.5	80613.3	636472.3.3	136917.6
Percent of Total output	100%	7.16%	92.84%	10.44%	82.40%	17.60%

Source : R.S. Dube and K.P. Misra, *op. cit.*, 1984, p. 58.

TABLE 9.6

Shajapur District : Food Balance Sheet

Crops	Production	Seed	Gross food	Extraction	Net Food	Net loss
Wheat	41400	6136.0	75264.0	7526.0	67737.6	13662.4
Rice	3700	300.6	3399.9	339.9	3059.5	650.5
Jowar	116300	1529.7	154779.2	8033.0	106736.3	9563.7
Bajra	400	3.3	396.7	23.8	372.9	27.1
Barley	400	24.0	370.0	63.9	312.1	87.9
Maize	19700	249.0	19450.4	1361.5	18088.9	1611.1
Kodon	—	—	—	—	—	—
Gram	28900	3376.5	25423.5	5104.7	20418.8	8481.2
Masur	300	154.0	346.0	51.9	294.1	205.9
Urd	1000	45.9	954.1	124.0	830.1	169.9
Mung	5000	225.3	4774.7	620.7	4154.0	846.0
Arhar	9800	234.9	9666.1	1721.7	7843.4	1956.6
Teora	60	7.5	52.5	9.4	43.1	16.9
Groundnut	20925	3476.0	17449.0	9073.5	8375.5	12549.5
Mustard	—	—	—	—	--	—
Sesamum	100	3.4	96.6	56.7	40.6	59.4
Linseed	700	53.5	646.5	407.3	239.2	460.8
Total	288805	15821.9	273060.8	34518.8	247545.9	52339.0
Per cent of Total output	100%	5.50%	94.52%	11.95%	85.69%	17.43%

Source : R.S. Dube and R.P. Misra; *op. cit.*, 1984, p. 59.

foodgrains in India. This would suggest that the rate of loss will be smaller in a region where cereals dominate the cropping activity than that in a region where pulses and oilseeds have a predominantly greater share in farming and vice-versa. Consequently, the average percentage of loss of grains from all sources comes out about 20.39 in Madhya Pradesh, which is higher than the average rate of loss suggested by the previous workers.

The unsuitability of the use of average rates of losses have been further substantiated from the food balance sheets of the wheat-gram producing Sagar district, the rice-teora growing Raipur district, and the jowar-wheat producing Shajapur district (Tables 9.4, 9.5, 9.6). The rates of total loss have been worked out at about 24 per cent in Sagar, 20 per cent in Raipur and about 17 per cent in Shajapur. These facts would disprove the employment of "an average rate of loss approach" for the estimation of foodgrain availability in a region.

The Elements of Food Sufficiency

Sufficiency of food products in a territory is gauged in the context of standard dietary requirement of an individual. There are three approaches :

1. Comparision of per capita availability of foodstuffs with a balanced dietary composition of an average individual.
2. Conversion of per capita availability of foodstuffs into food energy (calories) and its comparison of with the standard requirement (2400 calories) of an average individual and ;
3. Conversion of per capita availability of food stuffs into various nutritive factors and their comparison with their standard requirement of an average person.

The requirement of different foodstuffs for a balanced vegetarian diet of an average Indian is appended in Table 9.7. The available per capita amount of different food products may be compared with their requirement in a balanced diet in order to ascertain the level of their sufficiency in the territory under question. These per capita amounts of food products may be converted into calories and nutritional factors and their sufficiency may be judged by comparing them with their standard requirement for an average individual as summarised in Table 9.8. If available amount of food stuffs or nutritional value derived therefrom is less than the dietary recommendations, the area in question would be marked as deficit in regard to the stuff or factors under examination and if food production is more than the recommended amount, the area would be considered as surplus. These surpluses-deficit pictures would be helpful in the formulation of a rational food distribution policies.

TABLE 9.7

Composition of Balanced Vegetarian Diet of an Average Indian Individual

Foodstuffs	*Amount (grams)*
1. Cereals	369.5
2. Pulses	68.6
3. Leafy vegetables	107.4
4. Other vegetables	124.5
5. Fruits	37.1
6. Milk	178.4
7. Fats & Oil	37.6
8. Sugar and Jaggery	46.3

Source : Gopalan, C. *et al.* (1971) : Diet Atlas of India. Hyderabad National Institute of Nutrition, pp. 13 and 101.

TABLE 9.8

Food Factors : Requirement of an Average Person in India

Food factors	*Standard Requirement*
1. Energy	2400 calories
2. Protein	68 grams
3. Fat	60 grams
4. Carbohydrate	605 grams
5. Calcium	900 milligrams
6. Phosphorus	1000 milligrams

7. Iron	17 milligrams
8. Carotene	1500 micrograms
9. Thiamin	1.0 milligram
10. Niacin	12.6 milligram
11. Riboflavin	1.4 milligrams
12. Vitamin-C	49 milligrams

Source : Noor Mohammad (1978) : Agricultural Land Use In India. Delhi : Inter India Publications, p. 183.

References

Ali Mohammad (1977) Food and Nutrition in India. New Delhi : K.B. Publications.

Akroyd, W.R. (1941) The Nutritive Value of Indian Foods and the Planning of Satisfactory Diets Health Bulletin, No. 23, pp. 25-27.

Chakravarti, A.K. (1970) "Foodgrain Sufficiency Pattern in India", Geographical Review, 60(2), pp. 208-228.

Directorate of Agriculture (1979) Agricultural Statistics of Madhya Pradesh, Bhopal : Agricultural Press.

Dube, R.S. "Agricultural Productivity in Madhya Pradesh" (unpublished).

Dube, R.S. and R.P. Misra (1984) "Evaluation of Foodgrain Availability in India : An Alternate Approach", Asian Geographer, Vol. 3, No. 1, 1984.

FAO (1949) : Handbook for the Preparation of Food Balance Sheet, Rome,

FAO (1957) State of Food and Agriculture, Rome, p. 72.

FAO (1974) State of Food and Agriculture, Rome, p. 135.

Gopalan, C. and *et al* (1971) Diet Atlas of India, New Delhi, Indian Council of Medical Research.

Gopalan, C. and *et al* (1980) Nutritive Value of Indian Foods, Hyderabad: National Institute of Nutrition.

Jawahrlal Nehru Agricultural University Jabalpur, Krishi Vishwa, Vol. 2,3 and 4, 1980.

Knight, H. (1954) Food Administration in India. Stanford : Food Research Institute, California, p. 25.

Misra. R.P. (1982) : Population and Food Supply in Madhya Pradesh Unpublished Ph. D. Thesis, University of Saugar.

Noor Mohammad (1978) Agricultural Land Use in India, Delhi : Inter India Publications.

Shafi, M. (1960) "Measurement of Agricultural Efficiency in Uttar Pradesh", Econ. Geog., 36 (4), pp. 296-305.

Singh, B. (1951) Population and Food Planning in India, Bombay, pp. 11-12 and 85-86.

Stamp, L.D. (1958) "Measurement of Land Resources", Geographical Review 48(1), pp. 110-116.

CHAPTER TEN

DIMENSIONS OF FOODGRAIN SUFFICIENCY IN INDIA*

The level of foodgrain availability in India may be worked out by making use of the methodology discussed in the preceding sections of this context. The food balance sheet prepared for the country as a whole is appended in Table 10.1. A careful examination of the data summarised in this table would suggest that of the total foodgrain output, some 141 million tonnes would be lost owing to the various sources of losses arising from seed, cattle feed, extraction, etc. This would, then, give the per caput per day availability of some 481 grams of foodgrains—about 428 grams of cereals, 37 grams of pulses, and 16 grams of oilseeds. This would suggest, then that there is an overall excess of cereals as compared with their recommended requirement in the food consumption of an average Indian individual (Gopalan, 1971), but there is an acute deficiency of pulses and oilseeds.

The conversion of the per caput per day available food into food energy and food factors are recorded in Table 10 2, which informs deficiency of food energy (calories), proteins, fats, carbohydrates, calcium, vitamin A, Vitamin B_2 (riboflavin) and vitamin C, and surplus of phosphorus. iron, thiamin, and niacin in the Indian food supply. The supply of fats, calcium, vitamin A, vitamin C, and riboflavin from foodgrains is deficit by 61 to 94 per cent while deficiency of food energy (calories), proteins and carbohydrates is contained within the moderate range of 5 to 33 per cent. The supply of iron and niacin stands abundantly surplus against a moderate

* Based on R.S. Dube and R.P. Mishra, : "Evaluation of Foodgrain Availability in India : An Alternate Approach". Asian Geographer, Vol. 3, No. 1, 1984, pp. 58-67.

TABLE 10.1

India : Food Balance Sheet, 1981

Crops	*Production '000 kg*	*Seed '000 kg*	*Gross food '000 kg*	*Net food '000 kg*	*Per caput per year gm.*	*Per caput per day gm.*
Rice	53773400	13359.1	53760040	40384036	70.76	193.85
Jowar	11436100	2421.9	11433679	10633321	15.55	42.60
Bajra	5566400	1139.3	5565260	5231345	7.65	20.96
Maize	6199400	691,2	6198708	5764799	8,43	23.10
Small millet	5094300	710.2	5093589	3208961	4.69	12.86
Wheat	35507800	22641.2	3538519	31936643	46,70	127.96
Barley	2141700	1370.9	214029	1776473	2.60	7.12
Total cereals	119719100	—	—	106.734	—	428.45
Gram	5738800	5781.0	573309	4587415	6.71	18.38
Tur	1887400	39523.0	1847876	1515258.7	2.22	6.07
Other pulses	4556700	715627	3841072	3149679.0	4.61	12.62
Groundnut	6208700	74333	5465370	26233776	3.84	10.51

Sesamum	514100	7167.3	606932.7	212911.7	0.31	0.85
Mustard	1860100	21261.0	1838839	901931.1	1.32	3.61
Linseed	535100	56470.5	478629.5	177092.9	0.26	0.71
Total oilseeds	911800	—	—	3.914	—	15.68
Sugar	15734000	—	15734000	15261980	22.32	61.15
Potato	10133000	—	10133008	8613050	12.56	34.51

Source : R.S. Dube and R.P. M isra, *op. cit.*, 1984, p. 60.

TABLE 10.2

Nutrients Available in India (Per Caput Per Day), 1981

State	Food-energy (Calories)	Protein (Gms)	Fat (Gms)	Carbo-hydrates (Gms)	Calcium (Gms)	Phosphorus (MG)	Iron (MG)	Vitamin A Caro-tene (ug)	Vitamin B Thia-min (MG)	Vitamin B Ribo-flavin (MG)	Vitamin B Niacin (MG)	Vitamin C (MG)
(1)	(2)	(3)	(4)	(5)	(6)	(7)	(8)	(9)	(10)	(11)	(12)	(13)
Andhra	1860.09	39.75	6.55	410.85	82.71	911.19	19.07	88.143	0.72	0.41	10.42	0.02
Assam	1904.27	21.08	7.76	437.20	62.43	484.77	9.75	14.06	0.25	0.20	5.85	3.52
Bihar	1275.09	33.40	5.51	274.3	69.75	712.67	11.70	76.29	0.84	0.38	10.22	6.46
Gujarat	1850.81	37.35	75.48	255.28	122.13	941.76	21.75	238.43	1.14	0.68	9.61	2.44
Haryana	4854.23	147.24	36.59	977.13	500.43	3632.35	103.89	551.82	4 66	2.62	40.87	9.05
Himachal Pradesh	2134.41	70.15	16.73	433.86	170.76	1974.57	37.42	336.58	2.53	1.07	18.33	5.75
J & K	1776.28	48.76	16.54	363.934	95.93	1320.11	21.58	200.66	1.37	0.60	11.67	0.83
Karnataka	2086.48	45.78	30.55	407.74	115.02	1055.53	26.72	161.98	1.26	0 54	12.21	1.47
Kerala	447.26	8.73	0.90	100.91	14.10	201.88	3.93	1.70	0.08	0.08	2.38	0
M.P.	1999.95	63.27	20.17	390.30	180.23	1442.37	36.88	223.42	1.78	0.90	16.01	2.49
Manipur	1687.23	33.92	4.03	379.49	52.20	812.75	14.47	32.24	0.41	0.30	9.09	2.46
Maharashtra	1872.33	42.77	17.86	384.60	121.51	922.33	22.78	180.81	1.24	0.55	11.16	0.38
Meghalaya	1212.95	24.31	7.16	263.92	53.67	542.22	10.59	79.86	0.52	0.22	8.02	36.34
Nagaland	1301.3	22.26	9.79	282.09	44.65	523.94	8.91	44.83	0.35	0.19	5.84	7.17

Orissa	1955.23	48.40	9.19	419.59	113.63	1007.13	20.93	92.32	0.73	0.45	10.86	0.94
Punjab	[6554.84	188.37	40.42	1300.68	636.49	5197.83	147.49	512.30	6.47	3.69	60.70	12.16
Rajasthan	2153.84	74.23	35.93	383.68	230.82	1740.55	42.63	390.52	2.35	1.20	17.80	1.09
Tamil Nadu	1611.71	29.98	33.76	296.71	60.14	691.91	15.15	61.05	0.50	0.32	8.10	0.62
Tripura	1659.58	32.79	4.30	371.84	57.42	758.35	15.29	16.76	0.37	0.31	9.32	6.20
Uttar Pradesh	1841.18	33.34	22.76	376.06	102.11	710.07	14.96	151.73	0.83	0.40	8.53	15.98
West Bengal	1403.99	33.60	4.75	309.10	72.60	705.55	16.18	53.14	0.57	0.35	9.31	17.78
India	2028.81	48.91	23.48	404.71	141.93	1173.92	29.38	161.45	1.35	0.72	13,55	6.05
Recommended Requirement	2400.00	68.00	60.00	605.00	900.00	1000.00	17.00	1500.00	1.30	12.60	1.40	49.00

Source : R.S. Dube and R.P. Misra, *op cit.*, 1984, p 61

surplus phosphorus and thiamin. These imbalance in the composition of Indian food is bound to cause considerable health hazards emphasising the need for a rational well-conceived planning of food production and food regulation system in the country. This would necessitate, therefore, the knowledge of spatial variations in the nutrient availability patterns in different parts of the country as a prelude to the initiation of corrective measures for removing the inherent nutritional imbalances in the Indian food system.

SPATIAL DIMENSION OF FOOD AVAILABILITY PATTERNS

The spatial variations in the availability of all the individual nutritional factors of food have been mapped by making use of the data at the state level (see maps 9.1 to 9.4). The classes for mapping the spatial patterns have been determined by fixing the class boundaries in the form of percentage variation from the recommended requirement of a nutrient for an average Indian person as fixed by the Indian Council of Medical Research (Gopalan, C. 1971). The areas where availability is greater than the recommended requirement are classified under the surplus categories while those where it is lower than the recommended requirement are arranged under the deficient categories. Nutrient Availability Regions have been delineated on the basis of Kendall's composit scores for all the states as arranged in the Map 9.4. Four types of areas may be identified on this map :

1. Surplus Areas,
2. Moderately Deficient Areas,
3. Excessively Deficient Areas, and
4. Extremely Deficient Areas.

1. Surplus Areas

These areas include Punjab, Haryana, Rajasthan, and Himachal Pradesh in the northwestern part of the country. These states are surplus in terms of proteins, phosphorus, iron and vitamin B_1 (thiamin), while they are deficient in terms of fats, calcium, vitamin A (carotene), vitamin B_2 (riboflavin), vitamin B_2 (niacin), and vitamin C, in fact there is an overall deficiency of these nutrients in the whole country. Besides, the region is partly surplus in terms of food energy (calories) and carbohydrates. Punjab and Haryana are excessively.

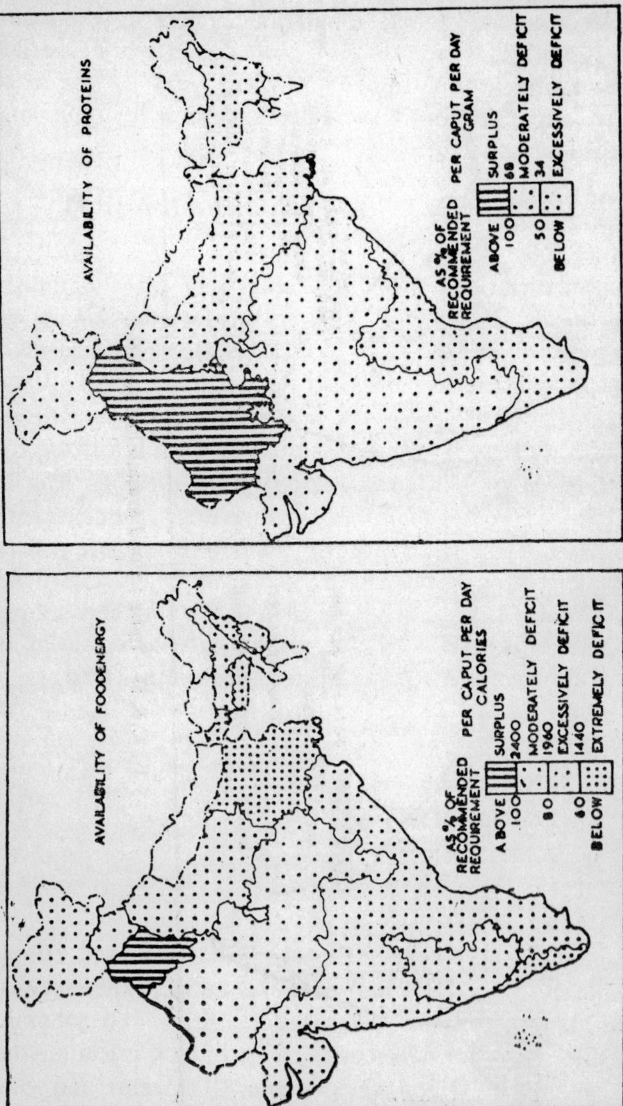

Map 9.1 : A & B : India : Food availability patterns.

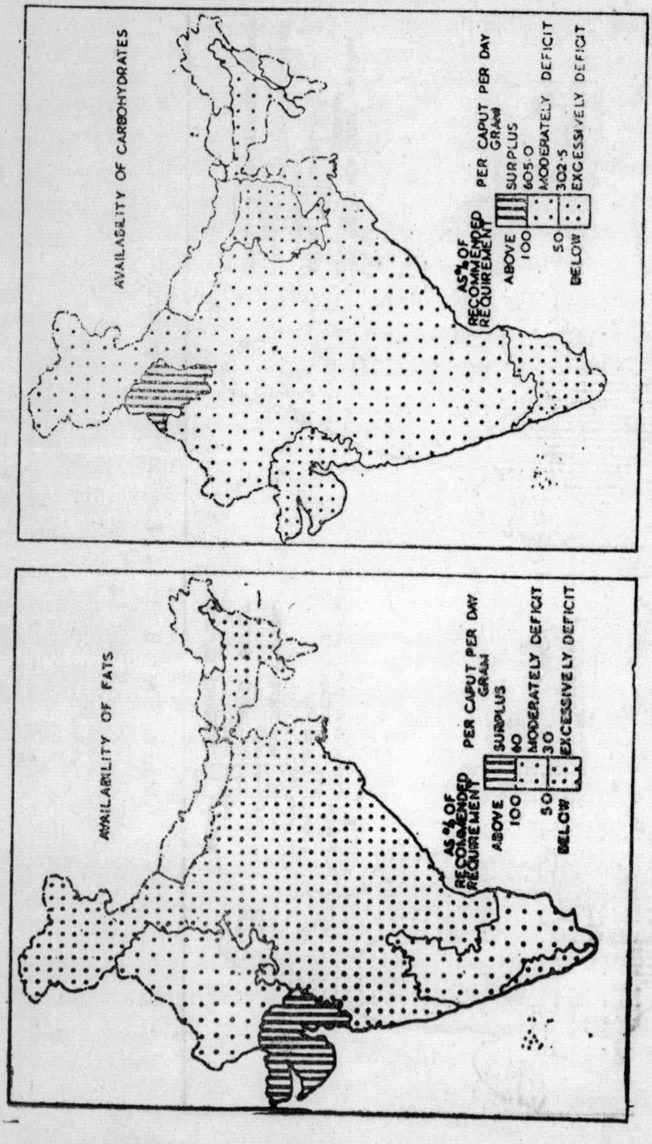

9.1 : C & D : India : Food availability patterns.

surplus in terms of proteins, phosphorus, iron and thiamin availability. Punjab seems to be more blessed in this respect than Haryana state. Rajasthan and Himachal pradesh provide surplus of proteins, phosphorous, iron and thiamin, but are moderately deficient in other nutrients.

Punjab stands, therefore, at the top in respect of food nutrition in the country. It is surplus by about 170 per cent from its indigenous requirement of food energy (calories) Table 10.2. The state is surplus in terms of proteins, carbohydrates, phosphorous, iron and vitamin B_1 (thiamin). It is moderately deficient in respect of fats, calcium, vitamin B_2 (riboflavin and niacin), and vitamin C. It is excessively deficient in the availability of vitamin A (cartone). Similarly, Haryana state is not much inferior to the Punjab state. It is surplus by a little more than 100 per cent from its recommended requirement of food energy (calories). The situation of other nutrients is the same as in Punjab state. Rajasthan and Himachal Pradesh have surplus of proteins, phosphorus, iron, and thiamin, but are moderately deficient in other nutrients. The availability of fats, calcium, iron and vitamin C in Himachal Pradesh is different from those in Rajasthan.

The high level of nutrient availability in the northwest Indian region (Map 9.4) is attributable primarily to the recent agricultural revolution and partly to the smaller size of population. The state supports some one tenth of the total population of the country, but it has 16.5 per cent of the total net sown area of the country. The area under foodcrops is about 80 per cent of the total cropped area and nearly 42 per cent of the gross cropped area is irrigated. In Punjab the share of irrigated area is, in fact, more than 81 per cent of the gross cropped area, which is highest in the country. The area under multiple cropping is little less than one third (30 per cent) of the gross cropped area. Consumption of fertilizers in this region is 43 kilograms per hectare, this figure is also highest for any Indian state. It follows, therefore, that the high nutrient availability in this region is attributable to the high foodcrop productivity and its high innovatory agricultural development.

2. Moderately Deficient Areas

These areas comprise a greater part of western and central India, including Gujarat, Madhya Pradesh, Maharashtra and Karnataka, covering about 27 per cent population and 40 per cent of

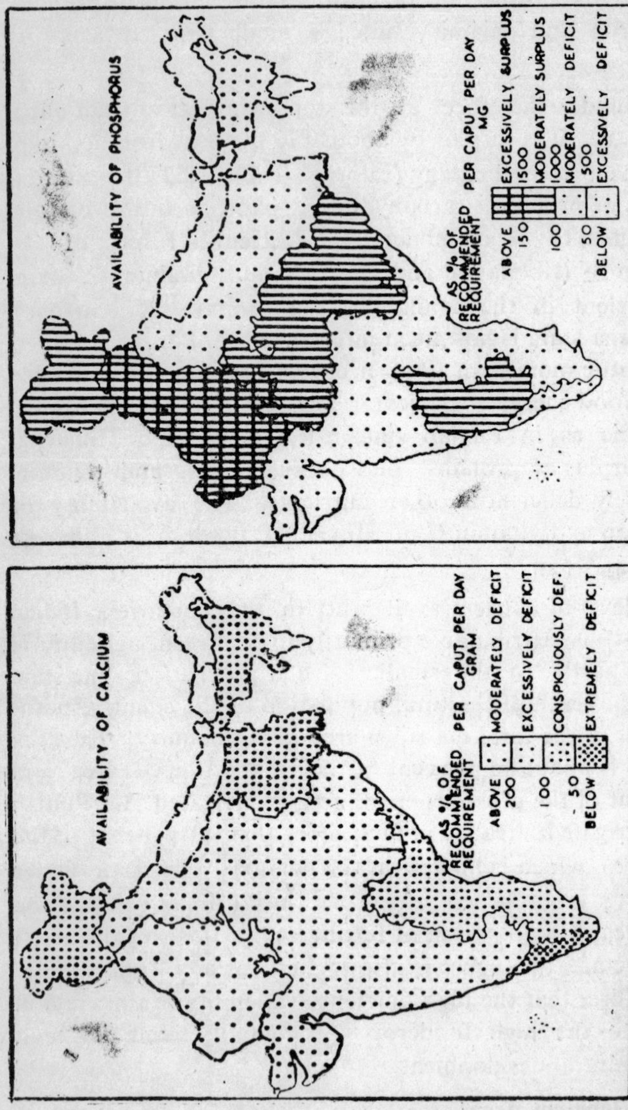

Map 9.2 : A & B : India : Food availability patterns.

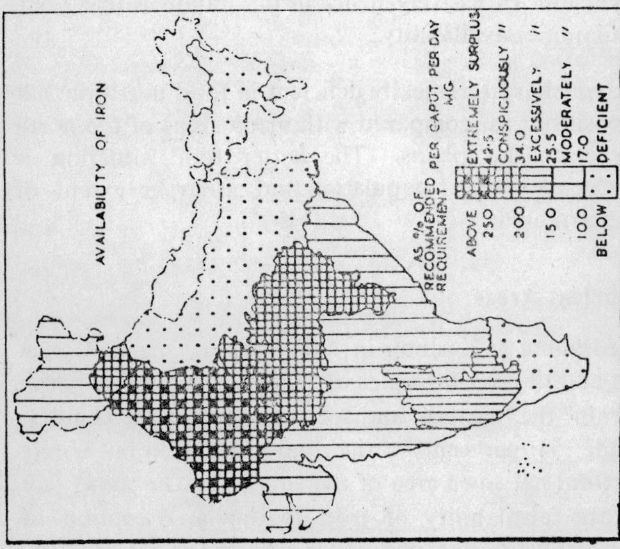

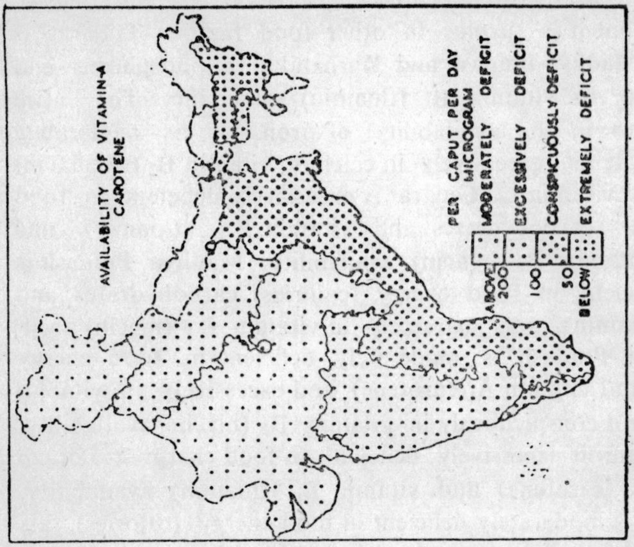

Map 10.2 : C & D : India : Food availability patterns.

Source : R.S. Dube and R.P. Misra (1984) : Evaluation Availability in India : An Alternate Approach, Asian Geographer (Hong Kong) Vol. 3 (1), pp. 51-67.

the net sown area of the country. This region is deficient in the availability of food energy (calories), proteins, carbohydrates, calcium, vitamin A (carotene), vitamin B_2 (riboflavin and niacin), and vitamin C, but it is surplus in other food factors. Gujarat is surplus in fats, Madhya Pradesh and Karnataka in phosphorus and Madhya Pradesh in vitamin B_1 (thiamin) availability. The entire region is surplus in the availability of iron, but is moderately deficient in proteins, excessively in calcium, vitamin B_2 (riboflavin) and extremely in vitamin C. Gujarat is excessively deficient in food energy (calories), carbohydrates and vitamin B_1 (thiamin), and extremely in vitamin B_1 (niacin) availability. Madhya Pradesh is moderately deficient in food energy (calories), carbohydrates and vitamin B_1 (thiamin), and extremely in vitamin B_1 (niacin) availability. Madhya Pradesh is moderately deficient in food energy (calories), fats, and vitamin A (carotene), and excessively in vitamin B_2 (riboflavin) and conspicuously in vitamin B_2 (niacin availability. Maharashtra state is excessively deficient in food energy (calories) fats, vitamin A (carotene) and vitamin B_1 (thiamin) availability. Karnataka state is moderately deficient in food energy (calories), fats and carbohydrates and excessively deficient in vitamin A (cartone), and vitamin B (thiamine) availability.

These areas are characteristically deficient in food nutrition but enjoy a better position as compared with many areas of the north India and southern coastal plains. The better food situation is attributable to the smaller size of population and a larger extent of the per caput net sown area.

3. Excessively Deficient Areas

They include Jammu & Kashmir in the northern part, Ganga valley in the north and the coastal states of Orissa, Andhra Pradesh and Tamil Nadu in the eastern and southern part of the country. These areas include 54 per cent of the total population but barely 39 per cent of the total net sown area of the country. The areas are usually surplus in the availability of iron (with the exception of Tamil Nadu). The surplus of phosphorous is reported in Orissa only. These areas are otherwise usually excessively to extremely deficient in food nutrition. They are one of the most problematic areas of the country in respect of food supply.

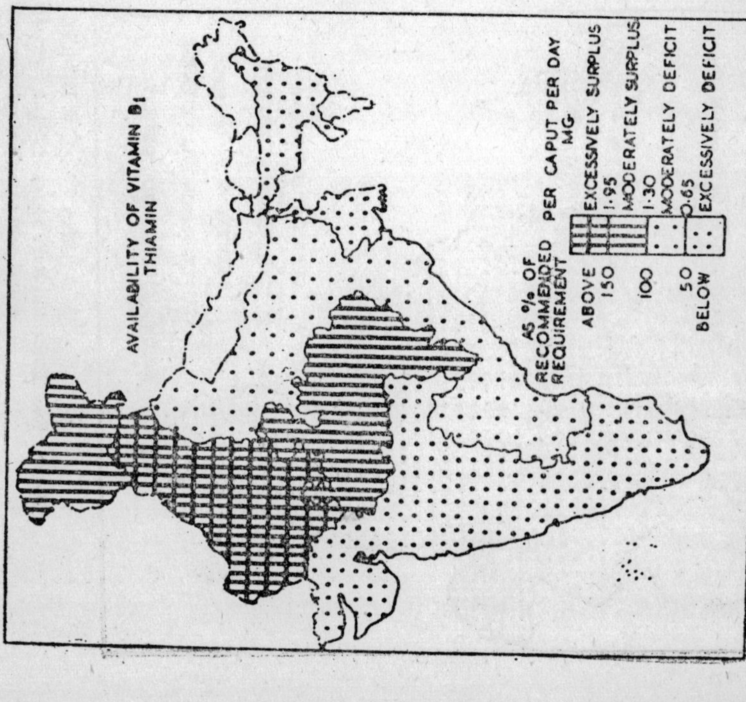

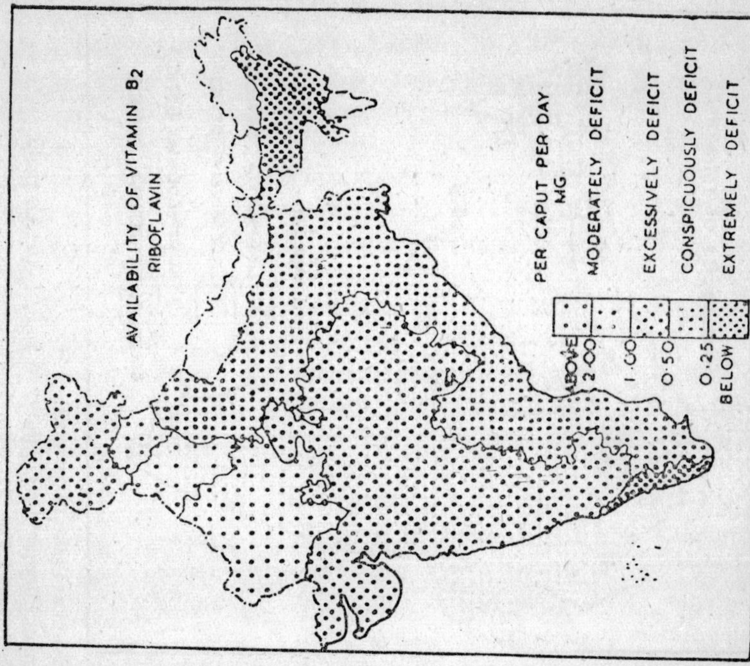

Map 9.3. A & B : India : Food availability patterns.

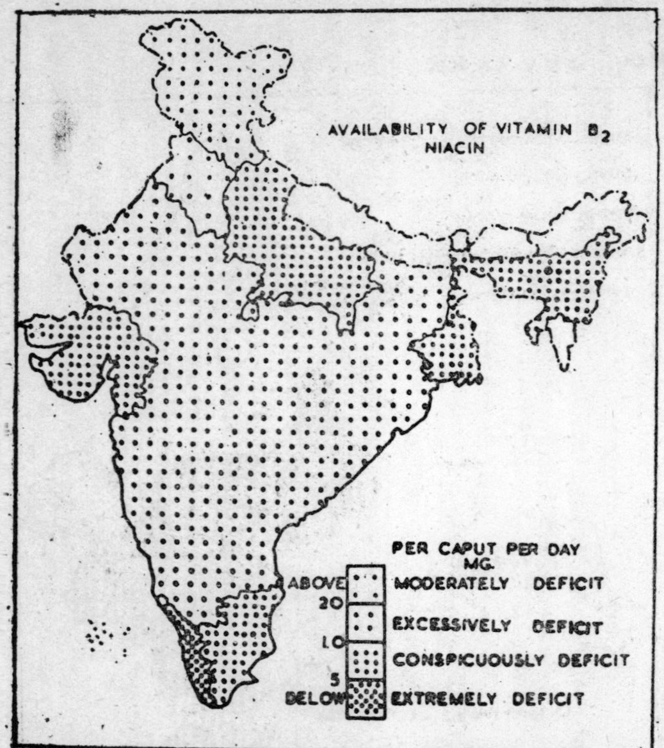

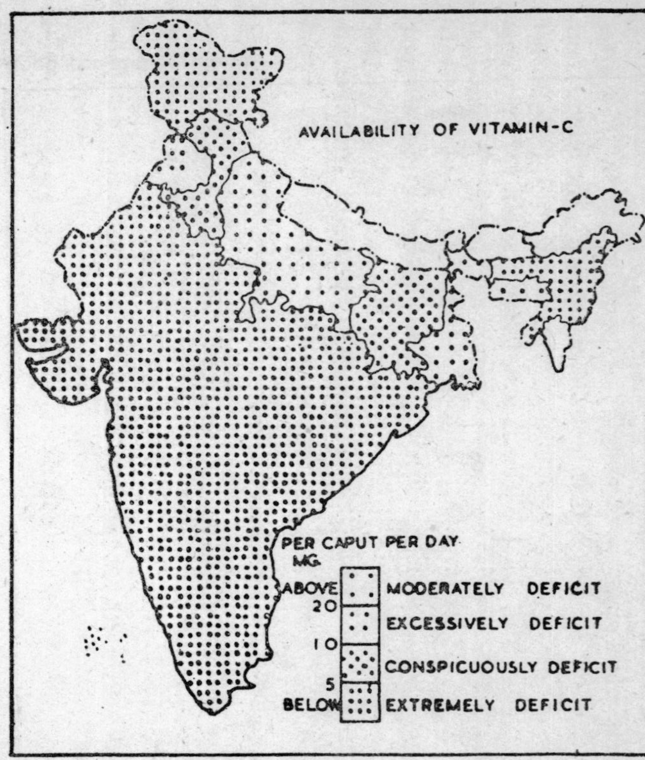

Map 9.3 : C & D India : Food availability patterns.

Source : R.S. Dube and R.P. Misra (1984) : Evaluation of Foodgrain Availability in India : An Alternate Approach/Asian Geographer (Hong Kong) Vol. 3(1), pp. 51-67.

The excessively low availability of food nutrition in these areas is attributable to the higher population densities, the hilly terrain, and inferior crops. For example, West Bengal (614 persons per square kilometer), Bihar (402 persons), Uttar Pradesh (377 persons) and Tamil Nadu (371 persons) are the states where population densities are highest in the country with the exception of Kerala where population density is 654 persons. The per caput cropped land is consequently 0.16 hectare in Jammu & Kashmir, 0.21 hectare in Uttar Pradesh, 0.17 hectare in Bihar, 0.14 hectare in West Bengal and 0.16 hectare in Tamil Nadu.

4. Extremely Deficient Areas

These areas comprise mainly the areas of north-eastern states. The state of Kerala is the only other state included under this category. These areas are excessively deficient in respect of proteins, fats, carbohydrates, phosphorous and vitamin B_1 (thiamin), while they are extremely deficient in terms of food energy (calories) calcium vitamin A (carotene), vetamin B_2 (riboflavin and niacin), and vitamin C availability. Iron is recorded moderate deficiency in this region.

Although the north eastern hill region is sparsely populated supporting about 47 persons per square kilometre, it is extremely deficient in food availability because of three factors :

(i) the smaller proportion of net sown area.

(ii) the predominance of a nutritively poor food crops system, and

(iii) the smaller cropping intensity, lower level in irrigation, fertilizer consumption etc.

The extreme deficiency situation in this part of the country would be appreciated when considered in the context of inferior level of agricultural development here. The net sown area is only 4 per cent of the country; its irrigated area is just 20 per cent of the gross cropped area; area under multiple cropping is less than 20 per cent and consumption of fertilizers is a little more than 10 kilograms per hectare. It follows, thus, that the hilly terrain and smaller agricultural development have combined to yield an extremely deficient nutrient supply in the region.

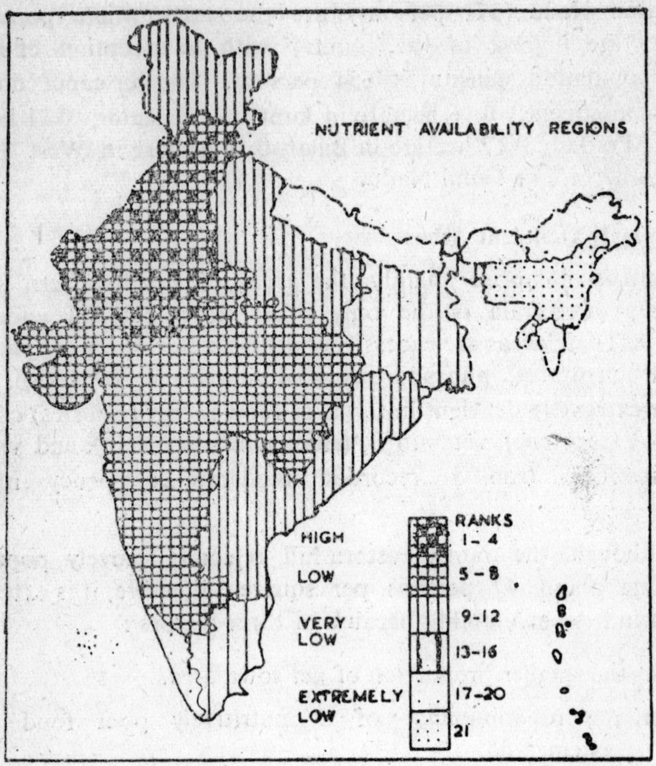

Map 9.4 : India : Food availability Regions

Source : R.S. Dube and R.P. Misra (1984) : Evaluation of Foodgrain availability in India: An Alternate Approach/Asian Geographer (Hong Kong) Vol. 3 (1). pp. 51-67.

It may be concluded, that greater availability of food nutrition in India is associated with areas where size of population is moderate and productivity of land is fairly high. Smaller availability is associated either with areas which support larger population or with those where the terrain is less hospitable for cultivation. The regional variation in the level of food supply in India is determined, therefore, by the combination of (i) nature of terrain, (ii) share of cultivated land, (iii) development of agricultural innovations, (iv) share of non-food crops, and (v) size of population.

5. Conclusions

1. The term foodgrain availability has been erroneously used by earlier workers for the net available food. It must be used for the amount of grains available for human consumption after undergoing losses of magnitude from extraction, destruction, seed and feed, ignoring the export-import factor of grain mobility.

2. Of the two approaches to the estimation of foodgrain losses, the one which measures the loss on seed, extraction and damages for individual grains at their individual rates must be preferred over the employment of some average of loss for all the crops, as has been done by many earlier workers. The seed and extraction rates differ so conspicuously from crop to crop that debiting of losses at an average rate from all the crops would yield a misleading estimate of availability.

3. India's food availability stands deficit in respect of food energy (calories) from foodgrains. Punjab and Haryana state in the north-western part of the country are the only exceptions. The less peopled Himachal Pradesh, Madhya Pradesh, Rajasthan and Karnataka are moderately deficient, while more crowded, Bihar, West Bengal and Kerala are extremely deficient in respect of food energy (calories).

4. There is a widespread deficiency of proteins, fats, carbohydrates, calcium, vitamin A, B_1, B_2 and C. A sizable part of the country possesses surplus only in respect of phosphorous and iron.

5. The northwest Indian region is placed in a relatively better position in respect of availability of food nutrition. It is deficient in fats, calcium, vitamin A, vitamin B (riboflavin and niacin) and vitamin C availability, but yields surplus of proteins, carbohydrates, phosphorous, iron and vitamin B_1 (thiamin) availability.

6. The Gangetic Plains in the north, the North-East Hill Region, and the southern and eastern coastal states are extremely deficient in food nutrition. The western and coastal Indian region is at a relatively better position in this respect.

7. The food nutritional availability level stems primarily from the level of population size, area under cultivation, and development of agricultural innovations. The north-west Indian region enjoys the benefits of higher productivity resulting from recent agricultural development. The greater size of population in the Gangetic plains and the coastal states of eastern aad southern India are responsible for acute deficiency of food nutrition. The lower proportion of cultivated area in Jammu & Kashmir or the smaller share of food crops in the Northeast Hill region and in Kerala contribute the smaller level of food availability there.

CHAPTER ELEVEN

NUTRITIONAL AVAILABILITY PATTERNS IN MADHYA PRADESH*

The study of nutritional availability patterns in a region assumes significance due to a number of reasons. It provides a reliable evaluation of food surplus or food deficit situation and affords an important tool for planning a rational food regulation system in an area. The surplus or deficit availability of nutrients affects the regional distribution of population and mobility behaviour of people and hence it may be employed as an yardstick for gauging the level of socio-economic attainments. The nutritional availability connotes the actual food available for human consumption in an area and in a given period. It must be distinguished from the food intake or food consumption and refers to the per caput amount of net food available for human consumption expressed in terms of the available quantity of the nutrients Since the purpose of nutritional availability is the identification of surplus or deficit conditions in respect of food production, the measurement of nutritional availability must rest on the determination of the quantity of food that may fetch human consumption locally or otherwise. An effort has been made in the present study to present a systematic and interpretative analysis of per caput food availability in different parts of Madhya Pradesh measuring in terms of the nutritive value of different food constituents. The aspect of nutritional availability, i.e. quantitative and qualitative, has been considered. Therefore, former has been measured as caloric content of the foodstuffs and the latter as the value of different nutrients, viz. proteins, vitamins, minerals etc. which are available in different foodstuffs.

---

*Reprinted from Trans. Inst. Indian Geog. Vol. 6(2) pp. 79-89 (Joint authorship with R.P. Mishra).

Methodology

FAO (1957) and Sukhatme (1962) have suggested two methods of estimating the level of food availability or food consumption in an area, viz. (i) the food balance sheet method, and (ii) the consumer survey method. The consumer survey method is based on a representative sample of households in a region and ideally covers their total expenditure in a given period for food and other items, the quantitative of food purchased and consumed. This method is useful for smaller areas. Madhya Pradesh, the study area, is so large that this method is difficult to be used. The food availability has been calculated with the help of food balance sheet method which has been adopted by Dube and Misra (1981). The food balance sheet shows the estimated per caput per day availability of a foodstuff in an area in a given period. It is measured from the total production, adjusted for in and out movements in trade, for changes in stocks, and for any quantities used for animal feeding, seed, industrial production, or other purpose apart from food. When such commodity balances are available for all main foodstuffs the caloric value and content of other nutrients, viz. proteins, minerals, vitamins, etc. can be calculated for the diet. Thus for the analysis of regional and spatial variations of food availability in Madhya Pradesh districtwise food balance sheets have been prepared. The data derived from those sheets were compiled and mapped into choropleth system of mapping to depict spatial variations therein. The primary consideration in the regional studies in respect of the evaluation of *inter se* relationship between population and food output is the determination of the surplus-deficit situation of areas in respect of food production. The study involves the consideration of all the possible sources of food in Madhya Pradesh in the range of cereals, pulses, vegetables, oils, milk and all kinds of flesh foods. The food production, data have been obtained from Commissioner of Land Records, Gwalior. Data for the analysis of the current patterns have been averaged from the five years data for the period 1976-77 to 1980-81, which are nearer to the Census year, *i.e.* 1981.

Nutritional Requirements

In the planning of a diet, energy requirements *i.e.* caloric food requirements of the individuals need to be considered by the nutritional geographer. Besides, energy food also comprises such nutrients as proteins, fats, minerals, and vitamins which are essential for the body. Experimental work in nutrition have revealed that an adequate diet must contain sufficient quantities of some forty different chemical substances and that only one of these needs to be deficient for the remainder to be inadequate to support health (Carpenter, 1969). FAO and WHO (1975) have recommended the requirement of calories, proteins, and other selected nutrients based on the amount considered necessary to meet physiological needs and to ensure good health, assuming that energy requirement are met. Obviously, the nutritional requirements vary with age, size of the body, sex, activity, occupation, foods to be digested, clothing, temperature of the body, the surrounding atmoshere, functioning of certain glands of internal decretion, etc. Mukherjee (1981), Swaminathan and Bhagwan (1960), Aykroyd (1962), and Gopalan (1980) have suggested the standard requirement of all the necessary nutrients in a daily diet as recorded in the following Table 11.1.

The first function of food is to provide the energy, which it supplies in a chemical form. The chemical energy (calories) is liberated in the tissues and converted by the muscles into chemical work which enables the individuals to carry out the physical, mental and socio-economic activity and provides heat necessary to maintain the body temperature. Generally proteins, fats, and carbohydrates are oxidised in the body to provide the energy required for various activities of life. While vitamins and minerals do not supply energy, they play an important role in the regulation of several essential metabolic processes in the body. Some minerals are important components in the body structures like bones and teeth. The deficiency of food nutrients causes a wide range of deficiency diseases and retards the working capacity of individuals.

TABLE 11.1

Recommended Per Caput Per Day Requirement of Nutrients

Nutrients	*Requirements*
Energy	2400.0 calories
Proteins	68.0 gram
Fats	60.0 gram
Carbohydrates	605.0 gram

Minerals :

Calcium	900.0 milligram
Phosphorous	1000.0 milligram
Iron	17.0 milligram

Vitamins :

Vitamin A (Carotene)	1500.0 microgram
Vitamin B (thiamin)	1.3 milligram
Vitamin B (riboflavin)	1.4 milligram
Vitamin B (niacin)	12.6 milligram
Vitamin C	49.0 milligram

Nutritional Availability Patterns

The quantitative and qualitative aspects of the nutritional availability have been considered in the present study. However, the former has been measured as the caloric content of the foodstuffs and the latter measures the value of different nutrients in different foodstuffs available for human ingestion. The quantitative food in developing regions such as Madhya Pradesh comes primarily from the grains. Other foodstuffs have negligible share in the daily diet of people in this State. On the other hand, the qualitative food availability depends upon the availability of fruits, fresh vegetables, milk, fats, and flesh foods. The availability of these foodstuffs is negligible in the study area. Table 11.2 reveals that Madhya Pradesh, as a whole, is deficient in the availability of calories, proteins, fats, carbohydrates, calcium, vitamin A, vitamin B complex, and vitamin C. Table 11.2 reveals that the deficiency of calories (19.78 per cent), proteins (7.63 per cent), and vitamin B (riboflavin, 23.57 per cent) is marginal, while the deficiency of vitamin A (68.42 per cent), vitamin C (88.45 per cent), fats (54.03 per cent), carbohydrates (40.85 per cent), and calcium (43.78) per cent) is substantial. On the other hand the state has recorded surplus availability of iron (64.65 per cent), phosphorous (43.08 per cent) and vitamin B (niacin, 11.59 per cent).

Caloric Availability

About one fifth of the districts of the state (22 per cent) are included under the surplus category where availability of calories is more than the recommended requirement *i. e.* 2400 calories per caput per day. Rest of the three-fourth (78 per cent) districts of the state are classed under deficient category. The surplus areas extend over the Malwa plateau in the western part of the state, the Raisen Vidisha plateau in the central part, and the Morena plateau and Lashkar plain in the extreme northern part of the State. The caloric availability in these areas ranges between 2716.37 calories in the Morena plateau to 3102.81 calories in the Umatwara plateau. On the other hand deficit areas extend over the Dandkaranya, the Chhattisgarh region and the Baghelkhand region in the southeast ; the Rewa plateau, and the Baghelkhand region in the north; the Central Madhya Bharat plateau and the Narmada valley in the central parts ; the Satpura in the south central part; and the Nimar uplands, and the Western Malwa plateau in the western part of the state. The caloric availability varies from 894 in the Bhopal plateau to 2340 in the Khichiwara plateau.

TABLE 11.2

Per Caput Per Day Availability of Nutrients in M.P., 1981

Nutrients	*Require-ment*	*Actual availa-bility*	*—Defici-ency +surplus*	*Percentage Departure*
Energy (calories)	2400.0	1965.27	—474.73	—19.78
Proteins (gram)	68.0	62.81	— 5.19	— 7.63
Fats (gram)	60.0	27.58	— 32.42	—54.03
Carbohydrates (gram)	605.0	357.86	—247.14	—40.85
Minerals :				
Calcium (mg)	900.0	506.01	—393.99	—43.78
Phosphorous (mg)	1000.0	1480.83	+480.83	+48.08
Iron (mg)	17.0	31.39	+ 14.39	+84.65
Vitamins :				
Carotene (microgram)	1500.0	548.64	— 95.36	—63.42
Thiamin (mg)	1.3	1.60	+ 0.30	+23.08
Niacin (mg)	12.6	14.06	+ 0.46	+11.59
Riboflavin (mg)	1.4	1.07	— 0.33	—23.57
Vitamin C (mg)	49.0	5.66	— 43.34	—88.45

Source : Calculated from the Nutritional Balance Sheet of Madhya Pradesh.

Proteins Availability

The suffciency of protein in a diet is an important measure of the adequacy and quality of the diet. Their importance is enhanced many times because they are required for the growth and development of body. It forms the machinery of the body, the pumps, muscles, catalytes, membrances and vital fluids like blood. Protein may be derived from two sources, i.e. animal sources and vegetable sources. Animal proteins are derived mostly from meat, fish, milk and its products. Among vegetable sources nuts and pulses are important. Soyabean is unique in vegetable protein sources while cereals and vegetables are poor in respect of proteins content.

In Madhya Pradesh proteins are obtained from cereals, pulses, and vegetables. The availability of animal proteins is considerably small. The per caput proteins availability is about 62.81 gram as against the recommended requirement of 68 gram, which is 7.63 per cent smaller than the prescribed requirements. In the state proteins availability varies from 33.87 gram in Bhopal district to more than 117.76 gram in Vidisha district. Spatially, north-eastern Malawa region in the west, and the Lashkar plain in the extreme north are the extremely surplus areas in regard to the per caput per day proteins availability, while the Morena plateau, the Chhatarpur upland, and the Narsinghpur plain are excessively surplus in this regard. On the other hand entire Satpura-Maikal range, the Rewa plateau, the Baghelkhand plateau, the Chhattisgarh region, and the Dandkaranya are the proteins deficit areas. In fact the rice growing areas of the state are deficit while wheat and jowar growing areas are surplus in the availability of proteins.

Fats Availability

The foodstuffs that are rich in fats are oilseeds and nuts. Cereals, pulses and vegetables contain negligible amount of fat. Oilseeds and milk are the only sources of fats in Madhya Pradesh. It has been found that the availability of fat is very low in the state and animal fats are practically negligible. Per caput per day availability of fats is only 27.28 gram as against the requirement of 60 gram which is 54.03 per cent deficit (Table 11.2). Fats availability is ten times higher in Rajgarh district (100.48 gram) than that of Raipur district (8.94 gram). Only three areas of the State, viz. the Morena plateau and the Gwalior plain in the north, and the eastern Umatwara plateau in the west, are surplus in terms of fats availability when compared with the recommended requirement. Rest of the state is deficit in this regard.

Carbohydrates Availability

They play a key role in providing dietary bulk, and thereby, contribute to bowel movement and elimination of toxic components from the system (Leveille, 1975). Carbohydrates can be absorbed into the blood stream and utilized by the tissue cells only as simple sugars, such as glucose. The availiability of carbohydrates is only 357.86 gram as against the recommended requirement of 605 grams. In fact, Madhya Pradesh, as a whole, is 40.8 per cent deficit in the availability of the carbohydrates. It is about three times higher in Shajapur district (546.64) than that of Bhopal district (157.26 gram). This reveals that entire state is deficit in this regard. Of those 84 per cent districts of the state are excessively deficient.

Minerals Availability

Minerals constitute an essential part of a perfect human diet. Three types of minerals are necessary in the diet, calcium, phosphorous, and iron. Bones and teeth are made up mainly of calcium and phosphorous, and iron is an important constituent for stimulation of blood formation.

Calcium gives strength and hardness in the teeth and bones. Besides, the contraction of the heart and muscles is activated by a proper intake of the calcium. Fortunately, milk, cheese, green leafy vegetables are sufficiently rich in this mineral. Many of the cereals contain some amount of this nutrient but rice is very deficient and therefore, insufficiency of calcium is one of the most important defects of the diet of rice eaters. The availability of this mineral is 506.01 milligram per caput per day registering 43.78 per cent deficiency. About 95 per cent districts of this state are deficit in this regard. Only two areas, viz. Morena plateau and the Chhatarpur upland are classed as surplus.

The utilization of calcium in the body is closely related to that of phosphorous because most of the calcium is deposited in the body either in the bones or teeth as calcium phosphate. Cereals, pulses, nuts and oilseeds are rich in phosphorous. Phosphorous deficiency is rarely encountered in the state because the food consumed by the population is predominantly cereal based. Madhya Pradesh is 48.08 per cent surplus in phosphorous availability. It ranges from 2843.54 milligram in Vidisha district to 821.28 in Bhopal district. About 91 per cent districts of the state are surplus in the availability of phosphorous.

Iron is an important factor in determining the capacity of the blood to transport oxygen to the tissues. Green leafy vegetables and millets especially bajra is good source of this nutrient. The state is excessively surplus in the availability of iron (84.65 per cent), when comparison made with the recommended requirement. It ranges from 14.8 gram in Surguja district to 75.48 gram in Vidisha district. Spatially, entire state is surplus in the availability of iron.

Vitamins Availability

They are essential for body growth and health. It has been seen that many diseases like night-blindness, dental decay, dental caries, scurvy, beri-beri etc. are chiefly the result of the deficiency of vitamins. Most of the vitamins are now being referred to in a majority of cases by names relating to their chemincal structures, as for example, carotene (vitamin A), thiamin (vitamin B), etc.

Vitamin A (carotene) is present in some animal products like butter, ghee, whole milk, curd, etc. Leafy vegetables such as spinach, amarnath leaves, etc. as well as ripe fruits such as mangoes, papayas and tomatoes are rich in this vitamin. The availability of carotene is about 548.64 microgram as against the recommended requirement of 1500 microgram, within the state. It varies from 148.99 microgram in Rajgarh district to more than 1005.89 microgram in Chhatarpur district. Spatially entire state is excessively deficit in the availability of this vitamin.

Vitamin B (thiamin) is concerned with the proper functioning of the nervous system and with carbohydrates metabolism. Its deficiency results in many diseases, viz. digestive disturbances, growth failure, weakness, anemia, beri-beri, etc. Among the commonly used foods the richest sources of thiamin are unmilled cereals, pulses and nuts particularly groundnut. The availability of thiamin is 1.6 milligram as against the recommended requirement of 1.3 milligram, suggesting a slight sufficiency of 23.08 per cent. It ranges between 0.52 milligram in Raigarh district of the Chhattisgarh basin in the east and 3.68 milligam in Vidisha district of the Malwa plateau. Nearly two-third state is surplus in the availability of this vitamin. These areas extend over the Madhya Bharat plateau, the Bundelkhand uplands, the Rewa plateau, the Malwa region, and the adjoining areas of the Satpuras.

Vitamin B (riboflavin) is concerned with several oxidation processes inside the cell. Deficiency of riboflavin in daily diet may cause loss of hair, soreness of the lips and tongue. Wheat, pulses, green leafy vegetables, milk, egg, fish are good sources of riboflavin. Availability of this vitamin in Madya Pradesh is 1.07 milligram as against the recommended requiremnt of 1.4 milligram, suggesting a deficiency of 23.57 per cent. It ranges between 0.48 milligram in rice growing Durg district of the Chhattisgarh basin and 2.27 milligram in wheat growing Vidisha district of the Malwa plateau. About 29 per cent districts of the state are surplus, while rest of the 71 per cent districts are deficit in the availability of riboflavin. Surplus areas extend over the Madhya Bharat plateau, the Bundelkhand upland in the northern part of the state; and the adjoining areas of the Malwa plateau in the western part.

Vitamin B (niacin) is a vitamin intimately concerned in several metabolic reactions. Deficiency of niacin causes pellagra disease. Cereals, pulses, nuts and meat are good sources of niacin and groundnut is particularly rich in this vitamin. The availability of the vitamin is 14.06 milligram as against the standard requirement of 12.6 milligram, suggesting a slight sufficiency of 11.59 per cent. It ranges from 7.41 in Jabalpur distict to 28.16 milligram in Vidisha district. Spatially, all parts in the south of the Narmada-Son valleys are deficient, while areas north of them are surplus in the availability of niacin.

Vitamin C is found in fresh fruits and green vegetables. Besides, orange, lemon, grape-fruits, and amla are richest sources. Madhya Pradesh is seriously deficient in the availability of this vitamin. The availiability of vitamin C is 5.56 milligram per caput per day against the requirement of 49.0 milligram. It ranges from 1.62 milligram in Durg district to more than 17.0 milligram in Indore district. Spatially, enrire state is excessively deficient in the availability of this vitamin. More than 90 per cent districts of the state are conspicuously deficient,

Nutrient Availability Regions

The spatial and regional disparities in the situation of nutrient availability in Madhya Pradesh will be more clear by the determination of nutrient availability regions. This regionalization has been based on the grouping of the districts by making use of the Kendall's ranking method (Kendall,1939). Districts are ranked in descending order for each of the indices and then, obtained ranks are summed and averaged to find composit score for each district. The first ranking distrist, has the lowest availability, and the last ranking district, has highest availability of nutrients. The regions are based on the availability of twelve nutrients, viz. calories, proteins, fats, carbohydrates, vitamin A (carotene), vitamin B (thiamin), vitamin B (riboflavin), vatimin B (niacin), vitamin C, calcium, phosphorous and iron. Three types of nutrient availability regions may be recognised, *i.e.* dynamic, prospective, and problematic.

1. Dynamic Availability Regions

They cover forty per cent districts including about one-third of the total population (31 per cent) of the state and extend in three compact areas, (i) the Malwa plateau in the west, (ii) the Bundelkhand upland in the north, and (iii) the Madhya Bharat plateau in the extreme northern part of the state.

The Malwa plateau is a fertile black soil tract and agriculturally developed part of the state. The nutritional availability is quite large because of smaller degree of population concentration and surplus production of foodstuffs. The density of population is lower than that of the Chhattisgarh basin. Proportion of net sown area in this region is very high, intensity of irrigation and double cropped area is moderate. Consequently, the high nutrient availability in these areas is associated with the small size of population, more food production, and more agricultural development.

The Madhya Bharat plateau is the second high nutrient availability region. Food production is higher than the size of population in this part, especially the Morena plateau yielded higher food production with larger population. The Lashkar plain and the Khichiwara plateau registered high food production and medium population size. As a whole this is one of the most developed parts of the state in terms of agriculture.

The Bundelkhand upland occupies Tikamgarh and Chhatarpur districts of the state. The Chhatarpur upland registered medium food production but smaller population, while Orchha upland (Tikamgarh) registered lower food production with smaller population size. These are the areas where the agricultural conditions are not different from those of the Madhya Bharat plateau.

2. Prospective Availability Regions

Nearly one-fifth districts supporting about 16 per cent population of the state recorded moderate availability of nutrients. They include the Rewa plateau and the Damoh plateau in the north, the Gwalior plain of the Madhya Bharat plateau; the western Sondwara plateau of the Malwa region; the Hoshangabad plain of the Narmada valley; and the Betul plateau and the Maikal plateau of the Satpuras. These areas are those parts of the state where net sown area is small and the area under double cropping in large. The population size is small.

The Rewa plateau recorded high food production and medium population size, while Damoh plateau supports smaller population and low level of nutrient availability. The situation in the Betul plateau, the Panna range, and the Sondwara plateau is similar with the Damoh plateau. The Gwalior and Hoshangabad plains are the areas where population size and food production both are medium.

It, therefore, follows that the prospective nutrient availability regions generally have medium or smaller size of population. They yield low to medium amounts of food production.

3. Problematic Availability Regions

About forty per cent districts supporting about 53 per cent population of the state recorded lower availability of nutrients. They extend over the Dandkaranya, the Chhattisgarh region, the Baghelkhand and the Satpuras. These are those parts of the state where density of population region is highest and the cultivated area is small. The proportion of backward classes is high in these areas.

The rice growing tribal Dandkaranya in the extreme southern part of the state, recorded lower availability. It is associated with the very low production of cultivated area (only 19 per cent) and very high proportion of forested area (65 per cent). Area under double cropping, irrigation, high yielding varieties, and use of fertilizers is very low. These conditions resulted in lower food production and the level of nutrients availability is consequently low.

The Chhattisgarh basin, the rice bowl of the state, also recorded low level of availability. It is the most densely populated part of the state. The proportion of cultivated land is medium; and the double cropped area, irrigated area is high and very high respectively. But the larger size of population results in lower availability. Other conditions are not different from those of the Dandkaranya region.

The Baghelkhand, undeveloped region recorded very low proportion of net sown area, irrigated area, double cropped area, and lawer productivity of soils. These conditions resulted in lower food production and lower availability of nutrients.

The entire Satpura region except the Betul plateau also recorded lower availability of nutrients. The proportion of cultivated land, and double cropped area is very low. On the other hand size of the population is large. These conditions result in the lower availability of nutrients.

It, therefore, follows that the lower availability areas are generally those parts of the state where either the population size is very large or the agricultural development is low in comparison to the population size. Besides, the more urbanised parts of the state evide nce lower level of nutrient availability because of very low amount of food production and a larger size of population.

4. Conclusions

The quantitative and qualitative aspects of nutrients availability, have been discussed in the present study, are the best indicators to the determination of nutriem availability regions. The quantitative nutrient (calories) availability is about 1965 calories as against the recommended requirement of 2400 calories, suggesting a net deficiency of 18 per cent. About one-fifth of the total districts of the state are surplus in the availability of calories. They extend over the Kauntal plateau, western Umatwara plateau, northern part of the Malwa plateau, and the Raisen-Vidisha plateau. Nearly two-third districts of the state are deficient in this regard. They extend over the Dandakaranya, the Chhattisgarh region, the Baghelkhand region, the Central Madhya Bharat plateau, the Rewa plateau, the Bundelkhand uplands, the Narmada valley, and the Satpuras.

Madhya Pradesh is deficient in the availability of proteins, fats, carbohydrates, calcium, vitamin A, vitamin B_2, and vitamin C. It is surplus only in terms of the availability of iron, phosphorus, vitamin B, and vitamin B_2 (niacin). Nearly 1/3 population (31 per cent) of the state recorded surplus nutrient availability and about one-sixth (16 per cent) population recorded moderate deficiency of nutrients, while rest of the population (53 per cent) of the state is malnourished.

The areas in the north of the Narmada-Son valleys are surplus in the availability of nutrients and those in the south of the valleys are deficient in this regard. Moreover, the Malwa plateau, the Madhya Bharat plateau, and the Bundelkhand upland are the surplus areas of the state, while the Narmada valley, the Satpura Maikal region, the Baghelkhand region, the Chhattisgarh region, and the Rewa plateau are the deficient areas in respect of the nutrients availability.

It, therefore, follows that the rice growing eastern districts of the state are deficient and northern areas of the state are surplus. The Narmada-Son valley stands as a divide between the surplus north and the deficient south.

References

Aykroyd, W.R. (1962) : The Nutritive Value of Indian Foods and Planning of a Satisfactory Diet. Indian Council of Medical Research, New Delhi.

Carpenter, K.S. (1969) : "Man's Dietary Needs", in Hutchinson, J. (Eds.) Population and Food Supply, Cambridge, Cambridge University Press, p. 62.

Dube, R.S. and Misra, R.P. (1981) : "An Alternate Approach to the Evaluation of Foodgrain Availability". Paper was presented at the Third Indian Geography Congress, New Delhi, December 26-29 (Accepted for publication in Asian Geogr. Hong. Kong).

FAO (1957) : Caloric Requirement, Nutritional Studies No. 15, Rome, Food and Agricultural Organisation of the United Nations.

FAO and WHO (1975) : "Energy and Proteins Requirements", Food and Nutrition, Vol. 1-2.

Gopalan, C. et al (1980) : The Nutritive Value of Indian Foods, ICMR, New Delhi.

Kendall, M.G. (1939) : "The Geographic Distribution of Crops Productivity", Journal of Royal Statistical Society, 162 : 21-48.

Leveille, G.A. (1975) : "Issues in Human Nutrition and Their Probable Impact on Foods of Animal Origin", Journal of Animal Science, 41 : 733.

Misra, R.P. (1983) : Population and Food Supply in Madhya Pradesh (Unpublished Ph.D. Thesis).

—(1982) : "Nutritional Dimensions in a Rural Settlements in Madhya Pradesh". Submitted to Fourth Indian Geography Congress, Bombay, December 1982.

Mukherjee, R. (1931) : Food Planning for Four Hundred Million, London.

Sukhatme, P.V. (1962) : "The Food and Nutrition Situation in India—Part I" Journal of the Indian Society of Agricultural Statistics, 19 : 1-2. 49-87.

Swaminathan, Mand Bhagwan, R.K. (1960) : Our Food, Madras.

CHAPTER TWELVE

FOODCROP PRODUCTIVITY AND FOOD-GRAIN SUFFICIENCY IN AN OVER-POPULATED CENTRAL INDIAN TERRITORY*

The Rewa plateau in the northern part of the Indian table land between the over crowded Gangetic plains in the north and the less populated Central Indian highlands in the south affords an interesting case study in regard to the foodcrop productivity and foodgrain sufficiency. Though, the arithmatic density of population in this region (138 persons per square kilometre) would seem much smaller than that in many densely peopled parts of the Indian subcontinent, the pressure of population on available resources is indeed high. The mainstay of life is agriculture and the amount of per capita cultivated land (0.37 hectare) is quite small in the context of current subsistence economy of the country. Since the soils are poor and non-agricultural resources are sparse, the region faces the growing danger of hunger and mal-nutrition in future.

The over populous conditions in the region are also indicated by the diminishing size of land holdings which have been so much fragmented that about 56 per cent of them are smaller than two hectares and many are even smaller than 0.8 hectares (Table 12.1). Further, there is a consistent trend of emigration (Dube, 1979) indicating the diminishing capacity of the region to support its population. Lastly, the negative values of z-scores[1] for the Rewa

*Reprinted from Geographical Outlook, vol. XVI, 1980-81, pp. 87-94.

1. Z-scores have been calculated by considering the variables in all the districts of the state with the help of the formula :

$$z = \frac{X - \overline{X}}{S}$$ where z means z-scores, x means the variable. $\overline{x}$ means mean of the variable and s means the standard deviation.

plateau in respect of level of employment, degree of landlessness and availability of cultivated land, and the positive scores in respect of the intensity of cropping (calculated with reference to the average value for the state of Madhya Pradesh of which this region is administratively a part) substantiate that the region currently stands over-populated.

TABLE 12.1

Percentage of Provisional Operational Holdings in Rewa Plateau and Madhya Pradesh

Size in hectares	*Rewa plateau*	*Madhya Pradesh*
Less than 0.5	27.2%	20.2%
0.5-1.0	13.5	12.6
1.0-2.0	15.7	16.7
2.0-3.0	10.3	11.3
3.0-4.0	7.0	8.3
4.0-5.0	5.4	6.6
5.0-10.0	12.2	15.0
10.0-20.0	6.4	7.0

Source : Director of Agriculture, M.P. (1973) :

The knowledge of foodcrop productivity and foodgrain sufficiency in this region may, therefore, interest the geographers and the policy planners in formulating a future strategy for the creation and regulation of food supplies in this region.

Methodology

The foodcrop productivity has been measured in terms of the value of crops per hectare of the cultivated area and in that of the Yang's crop yield index (Yang, 1965). The foodgrain sufficiency has been evaluated by calculating the per capita daily availability of foodgrains in terms of weight, calories and other food nutrients. The deficiency or sufficiency has been estimated in regard to the minimum daily requirement of an average person as a part of the balanced diet (Gopalan et al 1971).

The method of calculation of the daily availability of foodgrains in this paper has slightly deviated from those adopted by earlier workers. The main problem was the determination of the rate at which the individual grains are subjected to loss arising from wastage between farm and kitchen and their use for seed and feed. Chakravarti (1970) has made an allowance at the rate of 16.5 per cent for the wastage, seed and cattle feed. A local inquiry made by me has, however, shown that the seed rates and the extraction rates differ from crop to crop and that the total loss of foodgrains from wastage seed and feed for all the crops of the region adds upto 28.02 per cent. The seeds and extraction rates have been fixed according to the results averaged on the basis of interrogation of the number or farming households.

The net available foodgrains have, then, been converted into calories and food factors ana their sufficiency deficiency status have been evaluated in accordance with the requirement of nutrients laid down for a balanced diet (Gopalan *et al* 1971). The growth of food grain supply has been predicted and a strategy for future food supply has been suggested.

The Foodcrop Productivity

The region grows mainly wheat, jowar, kodo-kutki, gram, peas, rice, maize, barley, tur, masoor, urd, mung, moth, linseed, rape and mustard. Of these, the yields of only the first five crops are higher than the average yields in the entire state, suggesting that the productivity of a greater number of crops is lower than their correspond yields in the entire Madhya Pradesh state, making the food supply situation worse in a region (Table 12.2) where physiological density of population (236 persons per square kilometre of the cultivated area) is much higher than the average density for the entire state (165).

The foodcrop productivity measured in terms of the value of crops (Mitra and Mukherjee, 1980) and in that of the Yang's crop yield index (Yang, 1965) puts the Rewa Plateau into a negative position in relation to the other districts of the state (Dube, 1981). The productivity based on the value of crops comes about Rs. 545 per hectare against Rs. 872 in the entire state. The crop yield index (94.4 per cent) also places the region on the lower side of the state average. The inverse relation *inter se* productivity and population density seems strange if considered in view of the established opinion in favour of a direct relationship (Robinson *et al*, 1961) and it appears that in the Indian context the density does not obey much, the dictates of the productivity at the micro scale (Dube, 1980).

TABLE 12.2

Yield of Principal Crops (Kg/hectare)

Crops	*Rewa plateau*	*M.P.*	*Crops*	*Rewa plateau*	*M.P.*
Rice	450	893	Peas	415	256
Wheat	798	794	Masoor	246	377
Maize	927	1076	Urd	206	209
Jowar	652	616	Mung-moth	223	264
Barley	643	878	Linseed	177	217
Kodo-kutki	374	228	Rape and		
Gram	642	529	Mustard	184	296
Tur	502	638			

Source : Director of Agriculture, M.P. (1973).

Foodgrain Sufficiency

The food supply position seems still worse when foodcrop productivity of the region is considered in terms of the level of per capita availability of foodgrains. The region stands at the bottom of the rank order slabs of the districts of the state in respect of the value of crops per person (Rs. 215 against Rs. 370 in the entire state). The region yields about 425 thousand tonnes of the foodgrains. If wastage, seeds and cattle feeds are debited from the total production, the net foodgrain availability comes about 305 thousand tonnes or about 465 grams per day per person. Though, his availability of the grains seems sufficient as a part of the average balanced diet (Gopalan *et al*, 1971), it is, in fact, insufficient for a community where about 90 per cent of the diet rests on the grains. This places the region in a deficient status in respect of the availability of energy from foodgrains, which comes out only 1600 calories per persons per day against a minimum average requirement of 2400 calories.

The deficiency of foodgrains in the Rewa plateau is also suggested from the availability of food factors from the grains (Table 12.3).

TABLE 12.3

Rewa Plateau : Per capita Daily Availability of Food Factors

Factors	*Available Value*
Protein	61.06 grams
Fat	10.28 ,,
Minerals	9.29 ,,
Fibres	10.98 ,,
Carbohydrates	291.56 ,,
Calcium	174.37 Milligrams
Phosphorus	1298.87 ,,
Iron	23.58 ,,
Thiamin	2.44 ,,
Riboflavin	0.64 ,,
Niacin	18.47 ,,

Source : Calculated from production data of foodcrops, 1970-71.

The region seems to be deficient in respect of calcium, thiamin, riboflavin, fat and minerals, whereas the available supply of protein, iron and niacin is in excess over the recommended daily requirement. This would suggest that there is need for a balanced food production for creation of better and balanced nutritional intake for the people of this region. In fact, a greater part of the pulses is grown as cash crops, which, if debited from the net available foodgrains, would place the region still at the more deficient status in regard to the availability of calories from the grains.

The growth of hunger and the future food supplies in the Rewa Plateau

The intensity of hunger and under-nourishment is likely to grow with the rapid growth of population in this region. If population increases at the current rate of natural increase, it is bound to double by the end of this century, when we will have some four million people to seek their share from the available food supplies of the region. If we assume the foodcrop production to grow along the curve fitted into the curve of foodgrain production during the period 1951-1971, we find that the region would produce about 450 thousand tonnes of the foodgrains by the end of the century. An allowance made for wastage, seed and feed would give a net foodgrain supply of some 324 thousand tonnes making a net daily availability of some 235 grams per head. This would mean that the region may face an excessive shortage of foodgrains at the end of the century if productivity is not substantially enhanced. The hunger will grow and the Malthusianism will have its way on the density of the people of the country.

The region is, however, likely to increase its irrigation capacity with the completion of the Ban Sagar project[2] and the experience of agricultural development at the Tikamgarh district of the state may seem to suggest that the region has the potential of increasing the yield of wheat threefold.[3] If this becomes possible, the region would acquire the minimum requirement of foodgrains. If some food nutrients are obtained from the other sources of food also.

Conclusions

It may, then, be concluded that if foodgrain production increases in the direction of the growth curve of the grains during the 1951-71 period and the rate of natural increase of population

2. The Ban Sagar project is under construction of the Son river for irrigation and other purposes.
3. The Tikamgarh district in the northern part of M.P. was selected for Intensive Agriculture Area Programme by the State Govt. since then the district has registered considerable agricultural growth as evidenced in the current yield of crops (the yield of wheat, for example, is more than 1600 kilograms/hectares), extent of irrigated area (gross irrigated area 30 per cent of gross cropped area), and in the use of chemical fertilizers (about 20 kilograms of NPK per hectare of the cultivated area). The yield crops have consequently increased with the result that the yield of wheat in this district is now more than twice as high as in the Rewa district notwithstanding the fact that physical conditions are not much different there. If proposition of irrigated land of the Rewa district (about 5 per cent) and the extent of use of fertilizers (about 6 kilograms/hectare) can be enhanced, the region may attain self sufficiency in foodgrains in the near future.

remains unaltered, the Malthusianism may have its way over the density of the people of this region. It may, however, be hoped that the Tikamgarh experience of agricultural development may be repeated in the Rewa plateau with the completion of the Ban Sagar project when the region may have a threefold yield of wheat and other crops. The need of the hour is to take up a planned development of farming. The share of crops for cultivation may be distributed in such a proportion as to make the net output sufficiency for maintenance of the balanced food supply. Cultivations of fruit, vegetables and animal husbandry must invite the attention of the farmers.

References

Chakravarti, A.K. (1970) : "Foodgrain sufficiency patterns in India" Geog. Rev. Vol. 60, p. 217.

Director of Agriculture, M.P. (1973) : Agricultural Statistics, Bhopal.

Dube, R.S. (1979) : Population of the Rewa Plateau : A Geographical Analysis, Kanpur, Sahitya Ratnalaya.

..... (1981) : "Agricultural Productivity and Density or Population in Madhya Pradesh". (Unpublished).

Gopalan, C. et al (1971) : Diet Atlas of India. Hyderabad. National Institute of Nutrition.

..... (1980) : Nutritive value of Indian Foods. Hyderabad, National Institute of Nutrition.

FAO : State of Food and Agriculture, Rome.

Hammond, R.: Quantitative Techniques in Geography. Oxford et al (1974) Clarendon Press.

Mitra, et al (1980) : Population, Food and Land Inequality in India, 1971. Bombay, Allied Publishers.

Mohammad, A. (1977): Food and Nutrition in India. New Delhi, K.B. Publications.

Robinson, A.H. et al (1961) : "A Correlation Regression Analysis Applied to Rural Farm Population Densities in the Great Plains". Annals Assoc. Am. Geogr, Vol. 51, pp. 211-221.

Yang, W.Y. (1965) : Methods of Farm Management Investigations for Improving Farm Productivity. FAO, Agricultural Dev. Paper No. 80, Rome.

CHAPTER THIRTEEN

LAND UTILIZATION AND OVERCROWDING IN A RURAL SETTING*

The Lohandwar village (24° 37'N. Lat. and 81° 27'E. Long.) is located in the Huzur tahsil of the Rewa district, Madhya Pradesh India. It is situated in a north-south orientation on the left bank of the Mahana stream, a tributary of the Tons.

The land use, cropping pattern, and population of the village Lohandwar have been changing. The pressure of population on the agricultural land has been increasing partly because of the growing population and partly because of the occupation of land by the inhabitants of surrounded villages. An effort has been made in this paper to analyse the various factors responsible for the rapid growth of population and its consequences to be seen in the changes in the cropping pattern and in the emigration from the village records of earlier than 1975—5 are not available but the trend of change can be observed. The data on present land use pattern and population are based upon the Patwari Khasra and a house to house survey** respectively.

The revenue boundary of the village contains six well separated clusters of settlements. The land of the village is a well drained rolling plain and slopes (0—3°) due east to the Mahana. The soils are thick deposits of black and red loams. The latter being found in a narrow strip along the stream. The soil erosion is vigorously

*Based on R.S. Dube (1969) : "Land use Patterns and Population in the Lohandwar village, District Rewa, Madhya Pradesh, India, Indian Geog. Jour. Vol. XLIV (3-4), pp. 27-36.

**This paper was presented in a symposium on Rural Settlements organised by the Department of Geography, M.S. University of Baroda under the auspices of 21st International Geographical Congress, 1968.

active. Small dams have been constructed by erecting walls of earthen embarkments 5 to 10 feet high in order to prevent soil erosion and draining away of the soil fertility. The mean max. temperature for the cold months—December and January—ranges between 17.3°—22.6° c. and 40.0—42.3° (Rewa station for the hot months May and June. The average rainfall of the region is about 1118.5 m.m. (Rewa station) and more than 90 per cent of the rainfall is received during the rainy months of July, August and September. The rainfall is uncertain and inadequate for raising the crops successfully. In the absence of irrigation facilities the rabi crops are completely destroyed during the draught. Kharif crops are also destroyed if therein is not timely and adequate. Under such circumstances of the natural environment this village represents an example of primitive economy of the region based upon agriculture.

There are two main agricultural seasons ; the kharif or the season of summer crops ; and the rabi or the season of winter crops. The kharif crops include kodo, jowar, arhar, mung, urd (usually sown in combination), paddy, till, sawan and kokun. The rabi crops include wheat (also sown in combination with gram), linseed, lantil, barley, peas, tobacco, vegetables etc.

Located at a distance of about three kilometres from the Rewa Allahabad Road, one of the busy routes of the region, the village hardly indicates any influence of scientific development. The agricultural developments and techniques are primitive and standards of living is neither electrified nor are there any facilities of education, medication sanitation and communication.

A variety of castes combine to form a population of 993 persons living in 177 households. The brahmins and parihar rajputs comprise more than 51 per cent of the total population. Rest of the population is under the category of scheduled castes.

The frequency of literacy and educatedness is very low. More than 81 per cent of the total population are illiterate, about 13½ per cent are literate, about 4 per cent have passed the middle school examination, and only 1.62 per cent are matriculated or higher, The law frequency of literacy and education could be attributed to the fact that the educated segment of population moves to the urban areas either for higher studies or for employment, although the number of such population in small.

One of the most outstanding social features of the village is Purdah system among the women of higher classes. The women of scheduled castes work on the farms. Some of the scheduled castes, viz. chamar, kols and kumhars are still leading a life of slaves. They do not have their own land to work. They are the leading agricultural labourers of the village.

Land use patterns (1954-55 to 1967-68)

The various uses of the land during the year 1954-55, 1960-61, and 1967-68 have been summarised in the Table 13.2. It will be observed that during the year 1954-55 about 82 per cent of the land was under cultivation, about 10 per cent was under non-agricultural uses and a little less than 2 per cent was waste land. During 1960-61 about 81 per cent was under non-agricultural uses and about 2 per cent was waste land. In 1967-65 the corresponding uses of land are more than 82 per cent, about 9 per cent and about 2 per cent.

The most remarkable changes in the use of land between 1954-65 and 1967-68 are the changes in the area of earthen embarkments and settlements which have recorded a change of more than 18 per cent and 142 per cent from their area in 1954-55. The increasing area under embarkments is an indication to the fact that peasants are putting their day to day efforts to prevent the soil erosion and draining away of the soil fertility by eracting dams in which water is collected throughout the rainy season. The water which is collected in these dams does not allow the growth of grasses and decaying materials further adds to the fertility of soils. Similarly the increasing area under settlements indicates a greater expansion of settlements and households in the village.

The area under fallow lands and waste land has also very slightly increasing. The area under groves and water bodies parts of which have now been taken under cultivation have decreased to more than 97 per cent and 81 per cent from 1954-55 acreage.

The Map 13.1 and 13.2 show the general land use pattern: It will be observed that groves occupy a narrow strip along the Mahana because of suitable soils and nearness of the streams, which may supply irrigation water during the early stages of the plant growth of the mango trees dominating the areas of the groves. The allow lands usually occur in the marginal tracts near the streams and also in the higher slopes of the western parts of the village.

CROPPING PATTERN

Kharif season: Kodo and jowar usually sown in combination with the arhar and mung are the leading khariff crops. They occupied 28.59 per cent of the total area of the village 34.90 per cent of the net cropped land and abaut 31.72 per cent of the gross cultivated land during 1954-55. During 1960-61 the proportion of area under these crops changed to 29.37 per cent of the total area, 36.49 per cent of the net cropped land and 35.03 per cent of gross cropped land. During 1967-68 the change was still higher the corresponding proportions, being 37.33 per cent, 45.35 per cent and 41.27 per cent respectively. The change in area under these crops from 1954-55 to 1967-68 has been 130.56 per cent, suggesting that in the absence of manures and fertilizers the soils are getting exhausted.

The paddy cultivation occupies next place in importance. There has been no remarkable change in the area under this crop during the period 1954-55 to 1967-68. The change was only 115.38 per cent.

The most significant distributional feature of the kharif crops is that kodo and jowar dominate higher and inferior land of the western part while paddy is cultivated in the areas adjoining the settlements and the Mahana stream, where either good quality soils are found or nearer as they are from the settlements it is easy for the farmers to throw the cowdung and other waste materials to enrich the general fertility levels of the soils.

Rabi season : The leading rabi crop in the village is wheat sown alone and also in combination with gram. The other important crops are gram, linseed, lantle, peas, barley, tobacco, vegetables and mustard seed.

The rabi crops in this village have shown the most remarkable changes. Firstly, the variety of crops have been reduced about to half. The cultivation of peas, sugarcane, vegetables, mustard seed and tobacco has been abondoned altogether in 1967-68. Secondly, the area under gram and under wheat in combination with gram have considerably increased to 342.72 per cent and 158.77 per cent respectively from the period 1954-55 to 1967-68. Thirdly, area under wheat, linseed, lantle and barley has been reduced to 50.26 per cent, 51.17 per cent, 95.85 per cent and 94.19 per cent respectively.

A variety of factors have combined to cause these changes in the cropping pattern. Firstly, the increasing pressure of population on the land has compelled the farmers to give more emphasis on the cultivation of cereals. Secondly, the decreasing trend in the area under wheat cultivation is an indication of the fact how the gradual exhaustion of soil fertility, availability of rainfall and absence of irrigation facilities have combined to produce these changes. The emphasis is now on the combined cultivation of wheat and gram to ensure success even if the rainfall was lower and not timely. Thirdly, the area under linseed which was formerly a cash crop, has decreased because of the following causes :

(1) The linseed now no more stands as a profitable cash crop. Other rabi crops are equally cash paying.

(2) The linseed cultivation rapidly exhausts the soil fertility and the land which has been improved after a careful effort of years cannot be devoted for its cultivation.

(3) Linseed normally could be successfully grown on the land which has not been dammed so far. With the increasing proportion of agricultural land being dammed, the linseed is loosing its opportunities of being cultivated.

(4) The increasing pressure of population has necessitated the cultivation of cereals which may be equally cash paying in case their produce is in surplus.

Lastly, the abandonment of peas cultivation may be attributed to the following causes :

(1) The peas is grown relatively on an inferior and drier soil. The land under dams could not be utilised for the cultivation of this crop.

(2) The use of peas as a cereal has been responsible for a physical disorder, viz. Letharsis among those peoples who were regular peas eaters.

POPULATION

The total population of the village in 1968 is 993.* The sex ratio is 998.58 males per thousand of females. Of the total population about 48 per cent are in the age group 0—15 or above 66 years (age of infancy, childhood, schooling and old age). Only 52.37 per cent are in the age group 15—65, which normally is the age of a working man in the village. The dependance of non-working population upon the working population is, therefore, very high. About 30 per cent of the total population of the village are cultivators, a little more than 19 per cent agricultural labourers and about 10 per cent are in Government service (more than 3 per cent artisans are also included in these categories). Thus, there are about 52 non-workers per hundred working persons producing a great economic burden on a working man. Of this non-working population about 30 per cent are absolutely dependant (i,e. non-working stage).

The Table No. 13.5 summarises the age wise marital status of the population. Interesting features of this table are that : (1) early marriage is now no more characteristic of the population of rural environment like that of this village ; (2) there are no unmarried females in the village, but there are 30 unmarried males aged 15—44, the per cent frequency of these unmarried males being 13.75 per hundred males ; (3) sex ratio of the widowed with age groups 15—44, 44-65 and above 65 is 30 per cent, 40 per cent and 44.4 per cent respectively. The lower sex ratio

* This figure refers to the period ending on 10th October, 1968.

could be attributed to the social and cultural characteristics prevailing in the village. A widowed woman of higher caste cannot remarry. The woman of lower castes can, however, marry for several times even if their previous husbands are alive.

About 33 per cent (324 persons) of the population was born outside the village indicating that they have immigrated from other places. The immigrants include women coming after marriage, labourers and cultivators who had their agricultural land in the village. 22 persons have emigrated from this village permanently and 130 persons (about 10 per cent of the present population) have gone out for earning their livelihood. The outgoing population, however, exclude those farmers who have migrated after being married to the bridegroom of some other locality. There are at least 14 families which give a precise information of their forefathers being migrated from the different villages of the Gangetic plains. These patterns of recent and past migrations, both internal and external are curiously interesting features of the population movement in this region.

Density of population

The changes in the crude Arithmatic density* of population in this village are given below which show that the pressure of population on the land is increasing at a very fast rate.

Crude Arithmatic Density

Year	*Persons per hectare*	*Hectare per head*
1911	0.85	1.17
1921	0.96	1.03
1931	0.99	1.00
1941	1.15	0.866
1961	1.39	0.716
1968 (Oct.)	1.74	0.575

* The present author uses the term crude arithmatic density to designate the density of population of the village in relation to its total revenue area.

The crude arithmatic density of population in 1968 (Oct) is 1.74 persons per hectare of land or 0.575 hectares (1.52 acres) of land per head of the resident population. Of the total area of 571.23 hectares, 112.12 hectares or 19.62 per cent is under the possession of outsiders. This would mean that the arithmatic density pressure of resident population must be calculated in regard to 459.13 hectares of land which is under the possession of the village population. The density of the population of the village on this land is 2.16 or 0.46 hectares of land per head. Of the total area of 459.13 hectares only 369.91 hectares was under cultivation during 1967-68. Assuming that there was no remarkable change in the cultivated area of 1868-69, the pressure of the population in 1968 on the cultivated area (agricultural density) is 2.68 persons per hectare or 0.368 hectares per head. This stage of situation is very alarming in view of the growing pressure of population on the land.

Population Growth

The decennial growth of population from 1911 onward has been shown in the following table—The table shows that the total increase from 1911 to 1968 has been 104.6 per cent with an average annual growth rate of 1.3 per cent. The growth of population has, however, not been uniform. The highest growth rate has been recorded during 1961-68, while the lowest during 1921-31. The low growth rate during 1921-31 was due to a widespread epidemic in the region. Assuming that the population grows at the rate of an

Year	*Population*	*Net increase*	*% increase*	*Total increase 1911-68*	*Average annual growth rate*
1911	485	—	—	—	
1921	551	66	13.60		
1931	570	19	3.48		
1941	659	89	15.61		
1951 (Projected)	724	65	9.86		
1961	795	71	9.80		
1968	993	198	24.92	104.6%	1.3%

average annual growth rate of 1.3 per cent, the population is bound to increase unless the measures of family planning programme are intensified.

Conclusions

This analysis of land utilisation, cropping pattern and population of the village reveals that a growing emphasis upon the food crop is due to the growing pressure of population on the agricultural land. The rapid increase in the population of this village has not only been due to the natural increase but immigration also has played a leading role in the population growth. Assuming that the present rate of population growth continues and the development of land resources remain static, it appears that the population of this village will have no alternative but to move to other areas. A tendency for such immigration is revealed by the recent trend of immigration from the village. It is, therefore, suggested that the Government should not rely upon the family planning programmes only. It must work for the development of the agricultural land by providing at least the irrigation facilities and fertilizers. It is good that the State Government is seriously interested in the construction of the Dimba Project in the region. Before the completion of the scheme the tube wells, and several streams of the region may be utilised for this purpose. In case of this village the Mahana and the Tanks may prove of great use if an electric line could be provided by the Government. If appropriate precautions are not taken, the people might face very disasterous consequence in near future.

TABLE 13.1

Cast Grp.	*Population*	*% of total pop.*	*Cast Grp.*	*Population*	*% of total population*
Brahmins	448	45.12	Kurmis	73	7.35
Parihar Rajputs	62	6.24	Chamars	136	13.70
Kachhis	42	4.22	Kolas	56	5.63
Carpenters	13	1.30	Kumhars	25	2.52
Iron-smith	27	2.71	Ahirs	8	0.80
Barbers	94	9.46	Kotwars	9	0.95

TABLE 13.2

Total Area of the village 572.23 hectares

Use of land	1954-55		1960-61		1967-68		Per cent change 1954-55-1967-68
	Area in hectare	% of total area	Area in hectare	% of total area	Area in hectare	% of total area	
Cultivated Land	467.92	81.92	459.65	40.87	470.05	82.29	100.45
Fallow land	37.05	6.49	33.95	6.98	40.87	7.12	109.88
Earthen Embark-ments	13.92	2.44	17.42	3.02	25.08	4.44	180.08
Settlements	5.81	1.02	5.80	1.02	8.25	1.44	142.02
Groves	26.19	4.59	29.49	5.16	7.27	1.27	97.23
Water bodies	10.92	1.91	10.85	1.90	8.86	1.54	81.27
Waste land	9.96	1.74	8.34	1.46	10.99	1.92	110.36

TABLE 13.3

Gross Cropped Land:

1954-55 : 515:09 hectares

1960-61 : 479.69 hectares

1967-68 : 881.14 hectares

Net Cropped Land :

1954-55 : 468.03 hectares

1960-61 : 460.50 hectares

1967-68 : 470.05 hectares

Crops	*1954-55*			*1960-61*			*1967-68*		
	% of total area	*% of net cropped land*	*% of gross cropped land*	*% of total area*	*% of net cropped land*	*% of gross cropped land*	*% of total area*	*% of net cropped land*	*% of gross cropped land*
Paddy	5.46	6.67	6.06	7.38	9.17	8.80	76.30	.66	6.97
Kodo	1.30	1.59	1.45	0.74	0.91	0.88	0.53	0.64	0.58
Kodo-Arhar	1.19	1.45	1.32	0.21	0.26	0.25	2.46	2.99	2.72
Kodo-jowar	24.74	30.20	27.44	25.54	31.74	30.47	32.05	38.94	35.43
Arhar etc.									
Jowar	—	—	—	0.18	0.24	0.23	0.32	0.39	0.36

Jowar-arhar	0.32	0.39	0.35	—	—	—	0.48	0.58	0.53
Jowar-kodo	1.04	1.27	1.16	2.56	3.18	3.05	1.49	1.81	1.65
Arhar	—	—	—	0.13	0.16	0.15	—	—	—
Urd	0.20	0.25	0.23	0.14	0.17	0.16	0.07	0.08	0.07
Til	0.32	0.39	0.35	—	—	—	0.82	0.99	0.90
Sugarcane	—	—	—	0.15	0.18	0.18	0.17	—	—
TOTAL	34.63	42.27	38.40	36.92	45.88	44.04	43.79	53.21	48.41

TABLE 13.4

Gross Cropped Land :

1954-55 : 515.02 hectares
1960-61 : 479.69 hectares
1967-68 : 881.14 hectares

Net Cropped Land :

1954-55 : 468.03 hectares
1960-61 : 460.50 hectares
1967-68 : 470.05 hectares

Crops	*1954-55*			*1960-61*			*1967-68*		
	% of total area	*% of net cropped land*	*% of gross cropped land*	*% of totol area*	*% of net cropped land*	*% of gross gropped land*	*% of total area*	*% of net cropped land*	*% of gross cropped land*
Wheat	25.64	31.30	28.43	9.52	11.84	11.36	12.89	15.66	14.25
Gram	2.45	2.99	2.72	3.75	2.18	2.09	8.40	10.20	9.28
Wheat-gram	12.11	14.78	13.43	24.54	30.50	29.28	19.23	23.36	21.26
Linseed	9.86	12.04	10.94	3.43	6.74	6.74	4.82	5.85	5.53
Lantle	2.80	3.42	3.11	3.43	4.27	4.10	1.16	1.41	1.28
Barley	0.21	0.26	0.24	0.79	0.98	0.94	0.12	0.15	0.14
Peas	2.41	2.91	2.68	1.16	1.44	1.38	—	—	—

Vegetables	0.04	0.07	0.04	0.08	0.10	0.099	—	—	—
Mustard seed	—	—	—	0.925	0.2	0.03	—	—	—
Tobacco	—	—	—	0.025	0.2	0.03	—	—	—
TOTAL	55.52	67.77	61.58	46.75	58.09	55.85	46.60	56.70	51.59

TABLE 13.5

	Total			Unmarried			Married			Widowed			Divorced/separated		
Age Groups	*Per.*	*M.*	*F.*	*Per.*	*M.*	*F.*	*Per.*	*M.*	*F.*	*Per.*	*M.*	*F.*	*Per.*	*M.*	*F.*
All ages	993	500	493	413	240	173	516	240	276	61	17	44	2	2	—
0-5	163	80	83	163	80	83	—	—	—	—	—	—	—	—	—
15-44	281	147	—	134	214	124	90	23	44						
15-44	338	195	193	193	33	33	340	157	183	13	3	10	2	2	—
44-65	132	61	71	2	2	—	94	48	46	35	10	25	1	1	—
Above 65	29	17	12	1	1	—	15	12	3	13	4	9	—	—	—

CHAPTER FOURTEEN

A RATIONAL APPROACH TO THE MARKETING OF FOODGRAINS IN MADHYA PRADESH

Madhya Pradesh is one of the important foodgrain producing states of India. It produced about 10.40 million tonnes of cereals and 2.01 million tonnes of pulses in 1980-81. A sizeable part of the grains output of the State is damaged for want of suitable storage and adequate quick marketing facilities. Since composition of food intake of the people is largely grain-oriented marketing of foodgrains constitutes one of the important items of trade and business activities in the State. An evaluation of the sufficiency patterns of foodgrains in the State may help in highlighting the need for adoption of a rational and well-conceived plan for the marketing and circulation of foodgrains in this part of the country.

Methodology

Since marketing of a commodity depends mainly on the level of its surplus-deficit situation in an area; main task in this paper is centred on the evaluation of sufficiency of foodgrain production in different districts of the State. The surplus-deficit of cereals and pulses has been estimated in the context of net annual requirement of cereals and pulses calculated from the recommended daily requirement of an average Indian individual (cereals 369.5 grams, pulses 68.6 grams). The allowance for seeds, cattle feed, damages and loss in extraction has been debited from the farm-output of different crops in order to find out the estimated quantity of cereals and pulses available for human consumption (Dube et al, 1984). The deficit-surplus situation of the grains in different districts of the State has, then, been estimated by calculating the difference between the amount available and the amount required.

A tentative scheme has been worked out for movement of grains from the surplus to the deficit parts. Since there are conspicuous regional differences in the food habits of peoples of the State, the districts have been grouped into 'Food Divisions' delimited on the basis of constituents and administrative convenience. The surplus-deficit positions in different 'food divisions' have been calculated in order to determine their potential for export/import of cereals and pulses. A network of the flow of farm surplus to the deficit divisions has been proposed for effecting a rational movement of trade in cereals and pulses.

Sufficiency Patterns of Cereals

The State produced about 10.4 million tonnes of cereals in the year 1980-81 (Table 14.1). If seeds, cattle feed, damages and loss in extraction are debited, about 7.73 million tonnes of cereals are available for human ingestion giving an overall per capita daily availability of 407.42 grams. This would give a daily per capita surplus of

TABLE 14.1

Madhya Pradesh : Production of Cereals and Pulses, 1981

Cereals/Pulses	*Production ('000 tonnes)*
Rice	4053
Jowar	1815
Bajra	121
Maize	713
Wheat	3143
Barley	236
Kodo-kutki	237
Others	83
Total of Cereals	10401

Gram	1063
Tur	304
Teora	221
Urd	175
Moong-Moth	58
Masoor	108
Others	82
Total pulses	2011

Source : Directorate of Agriculture, Madhya Pradesh, Agricultural Statistics, 1981.

37.92 grams or an overall net annual surplus of 720 thousand tonnes of cereals suggesting a fairly good scope for trading activities of cereals.

Regional variations in the sufficiency of cereals in different districts of the State have been shown in Map 14.1 and quantities of cereals by which different districts stand surplus or deficit are recorded in Table 14.2. It would be seen that twenty four districts are classed as surplus and twenty-one as deficit in respect of the availability of cereals in the State. The deficit zone makes a continuous extensive stretch from the western border of the State to its northern border encompassing parts of Malwa plateau, Narmada valley, Setpura-Maikal ranges, Bundelkhand and Rewa plateau. The districts lying on the northern and southern sides of this zone are classed as surplus.

It would be evident from Map 14.2 that six districts, Bhopal, Indore, Khargone, Chhindwara, Narsinghpur and Jabalpur require more than 40 thousand tonnes of cereals each to satisfy their local requirements of cereals. Indore, Narsinghpur and Jabalpur required even more than 80 thousand tonnes each. The requirement of other districts in the deficiency zone is less than 40 thousand tonnes a year. These statements suggest existence of a fairly great potential for import of cereals from the suplus districts.

The districts where surplus of cereals is greater than 40 thousand tonnes lie mainly in the Malwa and Wainganga-Chhattisgarh regions. The surplus areas of Malwa extend in the closer proximity of the deficiency zone where they may find easy marketing access for their surpluses. The surplus areas of the Chhattisgarh extend at a greater distance from the deficiency belt.

It is evident that deficiency belt of the State lies between the zones of Malwa-Madhya Bharat region in the north and Wainganga-Chhattisgarh in the south with fairly good transport links. The surplus cereals from the surplus districts on either sides of the deficient belt may be conveniently transported to this belt.

THE SUFFICIENCY OF PULSES

Madhya Pradesh produced about 2.01 million tonnes of pulses in 1980-81 (Table 14.4). If allowances are made for losses accruing from seed, cattle feed, damages and extraction, they provide about 1.28 million tonnes of the pulses available for human consumption giving a daily per capita availability of about 67.41 grams. This would give an annual deficit of about 22.59 thousand tonnes showing possibilities for import trade of pulses from other states of the country.

TABLE 14.2

Madhya Pradesh : Surplus-Deficit of Cereals and Pulses in Food Supply Divisions

Division	*District*	*Cereals (tonnes)*	*Pulses (tonnes)*
I. Gwalior	Morena	+ 3,377	+16,725
	Bhind	+ 45,393	+37,303
	Gwalior	— 7,725	— 4,213
	Shivpuri	+ 14,492	+ 638
	Guna	+ 44,385	+ 6,841
	Datia	+ 2,418	+10,470
	Total	+1,02,340	+67,764

II. Bhopal	Bhopal	— 64,778	— 7,329
	Sehore	+ 48,548	+13,086
	Raisen	+ 330,88	—43,331
	Vidisha	+ 59,811	+29,237
	Rajgarh	— 12,459	— 559
	Total	+ 64,310	— 8,896
III. Ujjain	Ujjain	+ 42,351	+ 8,179
	Mandsaur	+ 81,660	+34,226
	Ratlam	— 6,222	+ 9,368
	Dewas	+ 11,089	+ 4,361
	Shajapur	+ 83,821	+12,396
	Total	+2,12,899	+68,510
IV. Indore	Indore	— 81,566	— 6,050
	Dhar	+ I1,089	+ 9,690
	Jhabua	- 24,917	+ 1,026
	Khargone	— 61,490	—18,145
	Khandwa	— 38,163	— 9,372
	Total	—1,94,999	—22,851
V Jabalpur	Betul	— 1,91,938	+34,502
	Jabalpur	— 1,91,938	—19,624
	Balaghat	+ 64,998	—18,861
	Chhindwara	— 77,416	— 180
	Seoni	— 6,132	— 4,686
	Mandla	+ 3,872	—15,750
	Narsinghpur	— 81,629	+48,688
	Hoshangabad	— 10,433	+ 2,884
	Total	— 2,28,396	+26,973

VI Sagar	Sagar	— 3,074	+ 1,026
	Damoh	— 6,016	+ 500
	Panna	— 2,759	+ 1,338
	Tikamgarh	+ 38,855	— 4,227
	Chhatarpur	+ 18,028	+ 6,775
	Total	+ 45,034	+ 5,412
VII Rewa	Rewa	— 28,912	— 9,451
	Sidhi	— 16,237	— 2,386
	Satna	— 15,536	— 11,824
	Shahdol	— 13,245	— 24,391
	Total	— 73,930	— 48,052
VIII Chhattisgarh			
	Raipur	+1,96,465	— 42,751
	Durg	— 3,101	— 7,857
	Rajnandgaon	+ 42,912	— 5,026
	Bastar	+1,38,981	— 40,853
	Bilaspur	+ 78,354	— 29,760
	Surguja	+ 3,744	— 30,855
	Rajgarh	+ 61,191	— 27,843
	Total	+5,19,543	—1,84,945
	Madhya Pradesh	+ 4,47,001	— 96,085

Source : Directorate of Agriculture, Madhya Pradesh, Agricultural Statistics, 1981.

Twenty-one districts of the State are classed as surplus in respect of the production of pulses, while twenty-four are classed as deficit (Table 14.2). The surplus districts lie north of the Narmada-Ken rivers, while deficient districts, with a few exceptions, extend south of them (Map 14.3). Five districts (Bhind, Morena, Mandsaur,

Vidisha and Sagar) have registered yearly surpluses of more than 16 thousand tonnes, while the districts where yearly deficit is more than 16 thousand tonnes include ten districts (Raisen, Khargone, Jabalpur, Shahdol, Surguja, Bilaspur, Raigarh, Raipur, Balaghat and Bastar), suggesting great scope for inter-district movement of trade in pulse. Since surplus and deficit districts form extensive compact stretches, the task of collection of pulses from the surplus zone and disbursement to the deficit zone would seem quite convenient and east to ragulate. Further, there is good scope for import of pulses from other states of the country for covering their deficit in this State.

THE PROPOSED FOODGRAIN CIRCULATION SYSTEM

The circulation of foodgrains from surplus to the deficit areas may be handled by different agencies at three levels :

(i) Collection of surpluses from the local village markets to the Central Godowns of grains located at a selected divisional place hereafter referred to as 'Divisional Godowns'.

(ii) Transportation of grains from the surplus 'divisional godowns' to the deficit 'divisional godowns', from where they may be disbursed to the local marketing places.

(iii) Interstate flow of foodgrain trade.

The state may be divided into eight 'Food Divisions' :

1. *Gwalior* (Morena, Bhind, Gwalior, Shivpuri, Guna and Datia districts);

2. *Bhopal* (Bhopal, Sehore, Raisen, Vidisha, and Rajgarh districts);

3. *Ujjain* (Ujjain, Mandsaur, Ratlam, Dewas, and Shajapur districts);

4. *Indore* (Indore, Dhar, Jhabua, Khargone, Khandwa districts);

5. *Jabalpur* (Betul, Jabalpur, balaghat, Chhindwara, Seoni, Mandla, Narsinghpur and Hoshangabad districts);

6. *Sagar* (Sagar, Damoh, Panna, Tikamgarh and Chhatarpur districts);

7. *Rewa* (Rewa, Satna, Sidhi and Shahdol districts; and

8. *Chhattisgarh divisions* (Raipur, Durg, Rajnandgaon, Bastar, Bilaspur, Surguja and Raigarh districts). The flow of the grains inside a Food Division be at the hands of local trading agencies, hereafter termed as 'Primary Agencies'. The surplus available with the primary agencies may be eventually transported to the Divisional Godowns by agencies which may be designated as 'Secondary Agencies'. The grains from surplus divisional godowns may be distributed to the deficit divisional godowns by a third level of trading agencies, hereafter termed as 'Tertiary Agencies'. The task of disbursement of arrival of grains in the deficit divisional godowns may be taken up by the secondary agencies and primary agencies may facilitate their distribution to the local markets.

The surplus-deficit situation of cereals and pulses in different food divisions of the State has been recorded in Table 14.2. It would be evident from this Table that Gwalior, Bhopal, Ujjain, Sagar and Chhattisgarh divisions are surplus in respect of the availability of cereals; Rewa, Jabalpur and Indore divisions are deficit in cereals. In case ef pulses Gwalior, Ujjain, Jabalpur and Sagar divisions are classed as surplus; Bhopal, Indore, Rewa and Chhattisgarh divisions as deficit. The State registered an overall surplus of 447,000 tonnes of cereals and an overall deficit of 9.6 thousand tonnes of pulses.

The deficit of cereals in Indore may be covered by importing about 195,000 tonnes of cereals from the neighbouring Ujjain division leaving behind a surplus of 17.9, thousand tonnes there (Table 14.3). The deficit of cereals in Jabalpur and Rewa divisions may be satisfied from an import of 228.19 thousand tonnes and 73.93 thousand tonnes respectively from Chhattisgarh divisions, leaving there a surplus of 305.09 thousand tonnes (Table 14.3). Surpluses in Gwalior, Sagar, Bhopal, Ujjain and Chhattisgarh divisions would combine to provide 534.67 thousand tonnes of cereals available for exporting to other States.

Gwalior, Ujjain, Jabalpur and Sagar divisions have registered surpluses in respect of pulses. The surplus from these divisions may be transported to the Chhattisgarh, Indore, Rewa and Bhopal divisions as shown in Table 14.3. The deficiency of about 96.09 thousand tonnes may be covered by importing pulses from other states of the country.

It may seem necessary to suggest that the flow of grains at the primary level may be handled by private agencies. The collection and disbursement of grains at the secondary and tertiary levels may be kept at the control of government agencies. Thus, collection of grains of divisional godowns from the local markets; flow of grains among divisional godowns; and disbursement of grains to the local markets in the deficit zones may be handled by Govt. agencies in order to keep efficient control on the regulation of food distribution system.

It would be seen that flow of foodgrains at such a large scale would require a great infrastructural trading network and would necessitate considerable planning.

Conclusions

1. The preceding discussion of the foodgrain availability and marketing patterns in Madhya Pradesh would suggest that knowledge of surplus-deficit of cereals and pulses is essential for rational circulation of grains from surplus to deficit areas. The surplus-deficit situation may be estimated by circulating difference between net annual availability and net annual level of recommended requirement of grains in the State.

2. The State may be divided into eight Food Divisions. The flow of grains from the farm to the local markets may be handled by primary agencies. The surplus with the primary agencies may be transported to the Divisional Godowns by secondary agencies. The tertiary agencies may carry the surpluses from Divisional Godowns to other deficit Divisional Godowns. The Secondary agencies of the deficit divisions may take up the task of distributing the grains to the local market places.

3. Gwalior, Sagar, Bhopal, Ujjain and Chhattisgarh divisions are surplus in respect of cereals. The deficiency of cereals in Indore,

Jabalpur and Rewa divisions may be satisfied by import cereals from Ujjain and Chhattisgarh divisions. The state would still have a surplus of 534.67 thousand tonnes of cereals ready for exportation to other states of the country.

4. The deficiency of pulses in Chhattisgarh, Indore, Rewa and Bhopal divisions may be fulfilled from imports from Gwalior, Ujjain, Sagar and Jabalpur divisions, leaving still a deficit of some 96 thousand tonnes which will have to be covered from import from other states.

5. The flow of trade of foodgrains at such a large scale to such far-off areas would require an efficient network of primary, secondary and tertiary agencies. It may seem desirable that collection at divisional godowns and their *inter se* flow of grains may be handled by the government agencies. The primary and tertiary level marketing may be left at the hands of private agencies.

CHAPTER FIFTEEN

A RETROSPECT

The nature of Agricultural Geography is in a state of constant change and modification. The modern Agricultural Geography is not concerned merely with description of the nature of farming in different earth regions. Its principal objective is the identification of fundamental issues pertaining the farm enterprises and farm operations in a territory. The content of the preceding sections of this volume is concerned with some of the issues ranging from the problems related with the classification and mapping of agricultural statistics to those related with the dynamism of land-use systems, agricultural production and circulation of agricultural produces.

Agricultural Data

The success of any regional analysis of the farm enterprises and farm operations, depends primarily on the presentation of data in classified and mappable forms, for this enables us to order agricultural facts. The present classifications are often subjective and arbitrary, since natural classes are rare to occur and to be identified. The need for finding unanimity on the adoption of some universally agreed method of data classification and data mapping would have to be appreciated, so that variation of patterns in different regions of a country could be avoided and an integrated approach at least at the national level could be obtained. As stated in Chapter two, the procedure of classification and mapping must be based invariably on a universally agreed typical value, which may be national average of any geographical variable or any other standard value identified for the purpose. If that is not possible simple arithmatic categories may be drawn, but the decision must rest on a well-conceived cartographic and statistical strategy.

It has been emphasised in Chapter two that agricultural data must be classed into groups on the basis of a universally agreed typical value and the class boundaries must be determined in the form of percentage variation from the typical value. Such a classification procedure would enable a uniform scheme of categories of groups of data in all the regional studies of a country. This would facilitate inter-regional comparability and cartographic ease in data presentation and would avoid the cartographic changes that predominates in the present regional studies of not only agricultural geography but also in other areas of Geography.

The Dynamic Approach

The dynamic approach to the understanding of the agricultural systems of the world regions often helps better appreciation of the characteristic features of agricultural activity. As emphasised and demonstrated in the Chapters three to six, the land use systems of the world have changed in a gradual succession from the primitive least productive forest fallow systems to the modern most productive multi-cropping systems. As such the nature and character of agricultural enterprises and farm operations in a territory must find better appreciation if they are studied as the components of the overall land-use cycle which has been in progression ever since the primitive phases of the evolution of agriculture in a territory. The agricultural aspects are often studied in isolation from their historical growth and the temporal changes that occur in the agricultural evolution of an area are very rarely taken into account in regional studies of agriculture. As such they fail to provide a complete understanding of the agricultural phenomena, its determinants and possible consequences in their correct historical perspectives.

It must be emphasised that like many other geographical attributes, agriculture also registers temporal changes in a cyclic progression. This has been highlighted by the research workers of the dynamic school of agricultural scientists. It may be pointed out, however, that agricultural workers have not visualised the course of events beyond the maximum intensity multi-cropping phases perhaps, because empirical observations have been difficult to find in different earth regions regarding the nature of the events that may eventually follow this last phase of agricultural evolution in a territory. It must be appreciated that a gradual retrogation of land-use or a sudden reversal of the agricultural system may be brought about by a set of multifarious factors. For example, depletion of labour or exhaustion of plant nutrients from the farm soils in the absence of any immediate technological innovations may lead to a gradual retrogation of land use pattern as highlighted in Chapters three and four. The destruction of a landuse system brought about by any physical or human calamity may cause the complete reversal of an agricultural system leading it to turn back to the previous primitive character. It must be visualised that all the multi-cropping maximum productive systems of the world eventually turn back either to the primitive forest fallow cultivation system or to any other subordinate cultivation system. After all my Malthusian doomsday is destined to cause inversion of land use progression.

Any system of land use of an area must hence be viewed as a part of the long history of change of agriculture. In fact many features of agricultural systems may find proper explanation only if they are considered in their broader historical context.

It must be appreciated that the longer cycle of land-use change would comprise numerous smaller cycles which would cause periodic changes and fluctuations in the agricultural characteristics. This has been demonstrated in the land use changes that have occurred in Madhya Pradesh as shown in Chapters four, five and six. These features of land use changes are indeed revealing since they exhibit similarity with the course of changes in many other natural and geographical elements.

Productivity, Sufficiency and Marketing of Agricultural Products

The study of Agricultural Geography of an area finds relevance only when it is observed in relation with the human beings inhabiting that area, for the ultimate destination of agricultural products is the human consumption. The agricultural geographer, therefore, finds interests in the evaluation of the spatial and temporal trends of agricultural productivity and sufficiency of farm produces. The temporal trends help in the understanding of improvement of deterioration of farming efficiency while the spatial aspects show the course of change that has occurred during the past periods, on the one hand, and highlight the contemporary problems of different areas on the other hand. The agricultural geographers have often analysed the areal variations in agriculture only for obtaining the understanding of the problems occurring in different parts of a territory. They have very rarely used spatial dimensions of agriculture as tools for the understanding of past changes in agriculture that may be lying preserved in the present land use systems of different parts of a territory. The facts compiled in Chapters seven to thirteen, highlight these aspects of agricultural geography. These chapters include discussions not only on some of the methodological issues of agricultural productivity and food sufficiency, they also provide information on various spatial facets of productivity and food sufficiency which help in the understanding of the course of historical changes which have occurred therein.

The temporal relationships between population and agricultural resources have been examined at micro scale in the Chapters eight, eleven and twelve. It may be noted that productivity-changes show a rythmic relation with population increase. So does the efficiency of food supply system. These observations have helped in examining the validity of the theoretical pronouncements advanced by the exponents of Malthusian and Boserupean schools. It has been concluded from the micro scale observations contained in this volume that Malthusian and Boserupean forces operate intermittently during the long course of population resource dynamism.

INDEX